MUNDANIA

How and Where Technologies
Are Made Ordinary

Robert Willim

T0256118

BRISTOL
UNIVERSITY
PRESS

First published in Great Britain in 2024 by

Bristol University Press
University of Bristol
1–9 Old Park Hill
Bristol
BS2 8BB
UK
t: +44 (0)117 374 6645
e: bup-info@bristol.ac.uk

Details of international sales and distribution partners are available at bristoluniversitypress.co.uk

© Bristol University Press 2024

British Library Cataloguing in Publication Data
A catalogue record for this book is available from the British Library

ISBN 978-1-5292-2144-2 hardcover
ISBN 978-1-5292-2145-9 paperback
ISBN 978-1-5292-2146-6 ePub
ISBN 978-1-5292-2147-3 ePdf

The right of Robert Willim to be identified as author of this work has been asserted by him in accordance with the Copyright, Designs and Patents Act 1988.

All rights reserved: no part of this publication may be reproduced, stored in a retrieval system, or transmitted in any form or by any means, electronic, mechanical, photocopying, recording, or otherwise without the prior permission of Bristol University Press.

Every reasonable effort has been made to obtain permission to reproduce copyrighted material. If, however, anyone knows of an oversight, please contact the publisher.

The statements and opinions contained within this publication are solely those of the author and not of the University of Bristol or Bristol University Press. The University of Bristol and Bristol University Press disclaim responsibility for any injury to persons or property resulting from any material published in this publication.

Bristol University Press works to counter discrimination on grounds of gender, race, disability, age and sexuality.

Cover design: Clifford Hayes
Front cover image: Robert Willim
Bristol University Press uses environmentally responsible print partners.
Printed and bound in Great Britain by CPI Group (UK) Ltd, Croydon, CR0 4YY

FSC
www.fsc.org
MIX
Paper | Supporting
responsible forestry
FSC® C013604

Contents

Preface

Mundania started to take shape for me some years into the new Millennium. Google was quite new, and so were so-called social media. There were no iPhones. Advanced AI-generated content was considered as science-fiction. Pervasive, sensor-laden, networked and computational systems were talked about, but they were far from being part of people's everyday lives. I was working with mixes of art and cultural analysis, and one part of this work was associated with site-specific art and something that at the time was called locative art. We used digital equipment such as GPS-receivers, to make artworks that questioned ideas about geography, tourism and experiences of places.

Some years later, all the disparate equipment that we had been using was bundled into small rectangular slates of glass and metal: smartphones. These devices were continuously connected to socio-techno-economic systems that became increasingly hard to outline or pinpoint. Devices and systems were used daily, but were never really domesticated, tamed or under control. Instead, technologies were mundanized: they became part of everyday lives but were impossible to grasp. Pervasive digitalization and so-called digital transformation have never been about the taming of technologies. Not domestication. Instead mundanization.

What if this expanding mundanization is part of the gradual emergence of the realm *Mundania*? A shape-shifting entity. Mundania is something that gradually arises, like an all-encompassing haze that gradually sets as layer after layer of ever more complex technologies and offerings are presented. I have not been able to get rid of that imaginary. What if most of us live in Mundania, or rather in various guises of Mundania? The realm where new technologies repeatedly become ordinary. Where the very ordinary gradually mutate. The atmosphere thickens, new layers crop up.

While dealing with Mundania, while writing this book, I have been working as a researcher and as an artist. I have been based at the Department of Arts and Cultural Sciences at Lund University, an interdisciplinary environment at the Faculties of Humanities and Theology. Here, studies of digital cultures meet and comingle with ethnology and aesthetic studies, with library and information studies and various kinds of media studies, as well as historically oriented research disciplines. It means that insights from digital cultures can be inspired by, for example, discussions and research within museum studies or musicology. This is the environment from which I have developed my ideas about Mundania. For some years I also worked at Halmstad University, in close collaboration with colleagues at RMIT and furthermore at Monash University in Melbourne. I also worked as an artist, exhibited and performed, parallel to my practices as an academic researcher, which extended the scope through which Mundania was evoked.

Most of the research leading to this book has been conducted in the project Connected Homes and Distant Infrastructures, financed by the Swedish Research Council (2017-00789). I have also worked within several interdisciplinary groups or networks, such as AI Lund, The Sound Environment Centre, CROCUS (Cultural Spaces and the Conditions of Creativity), crossdisciplinary research themes about The Digital Society as well as Search Engines and an advanced study group about The Seamless Life based at the Pufendorf Institute at Lund University. It is through manifold experiences stemming from these heterogeneous constellations and fields that this book is realized. Many thanks to all colleagues who have been part of these constellations. I also appreciate the help by Karin Gustavsson and Gabi Louisedotter at the Folklife Archives at Lund University who facilitated a directive about Connected Homes.

I want to especially thank all who have supported me in my work with Mundania, all those of you who have invited me for talks and presentations, who have commented, scrutinized and inspired during seminars, conference sessions, workshops, lectures and discussions. Magnus Andersson, Lars Gustaf Andersson, Kalle Åström, Stina Bengtsson, Martin Berg, Magnus Bodin, Dorothea Breier, Viktorija Ceginskas, Marie Cronqvist, Philip Dodds, Petter Duvander, Jessica Enevold, Tomas Errazuriz, Vaike Fors, Cecilia Fredriksson, Alison Gerber, Lizette Gradén, Ricardo Greene, Sanne Krogh Groth, Jutta Haider, Annette Hill, Melisa Duque Hurtado, Jenny Ingridsdotter, Anne Kaun, Eerika Koskinen-Koivisto, Stefan Larsson, Evelina Liliequist, Francisco Martínez, Gabriella Nilsson, Tom O'Dell, Sarah Pink, Jenni Rinne, Karin Salomonsson, Juan Sanin, Kim Silow Kallenberg, Olof Sundin, Kristin Veel, Julia Velkova, Kassandra Wellendorf, Jonas Wisbrant, and Lynn Åkesson, to just mention some of you. Many thanks to Asko Lehmuskallio and Orvar Löfgren, who read and commented on my manuscript. Nevertheless, I am of course myself responsible for what is in this volume, including any errors or discrepancies.

I am grateful to Bristol University Press who made this book possible, foremost Paul Stevens who believed in my proposal and who has been an inspiring companion in conversations about Mundania, academia and book publishing. Many thanks to the anonymous peer reviewers who scrutinized my work. Their constructive feedback was very valuable and greatly appreciated. Lastly, lovingly, thanks to my family, to Albert, Ida and Minna, who keep on reminding me that there is so much more in life than just books, and Mundania.

Robert Willim
Lund, June 2023

1

Arrival

What if everyday life was approached as a life in Mundania? A realm where more and more complex and incomprehensible technologies become part of the atmosphere. Layer after layer of these technologies is added. This realm shifts shape depending on context and the variabilities of everyday lives. Mundania emerges through the adoption of and adaptation to technologies that are gradually experienced as ordinary. It emerges through seductive offers and deals, built on the operations of a number of influential businesses. Mundania is then strengthened through recurring practices, rituals and routines through which technologies become intimately integrated in life. These circumstances make Mundania simultaneously banal and uncanny. Mundane and weird. At hand and ungraspable.

This book is a suggestion to imagine differently about so-called emerging technologies, about everyday life with foremost digitally engendered processes of computation bound together by vast technological and organizational networks and systems. It is also about continuously questioning what is experienced as and considered to be ordinary. What, when and for whom?

The book is based on a method I call 'probing'. It is part research, part artistic practice and a way to serendipitously move between and remix different cultural analytic concepts and perspectives (Willim 2017b; 2023). In this, I will juxtapose and mix thoughts about the most miniscule and overlooked phenomena with discussions on broader issues. I will weave these together with theoretical arguments and elaborations.

I will concentrate foremost on how technologies, in the form of concrete material and corporate operations and often clandestine arrangements, can come together in everyday life, among those using, or rather living through, the technologies. I suggest that the realm of Mundania is based on a process called 'mundanization'. I also suggest that Mundania is variable due to time, place and social circumstances. I will propose some variations of this realm and show how Mundania is based on processes of mundanization, and less so on domestication.

When technologies are introduced in Mundania, it is not about something wild becoming tamed and domesticated or converted step by step into controlled parts of everyday life. Instead, mundanization captures how complex arrangements of technologies and human organization maintain an incomprehensible unmanageability while still being transmuted into the ordinary, the mundane, the commonplace in people's everyday lives. In this sense, what emerges as the technologically ordinary in different contexts also houses ungraspability, unknowing and the potentially uncanny.

The empirical point of departure of the book is Sweden. It is a country that, around a decade into the second Millennium, was pictured and promoted as being one of the most technologically and digitally advanced nations in the world (Schwab 2019: ix). Geographically small, but with wider reach when it comes to technological development, Sweden is a Scandinavian and North European country of relative wealth, at the time of writing. A society with a high living standard and strong social security net that during the first decades of the second Millennium was gradually becoming increasingly inequal (Oxfam 2022).

An even more specific point of departure for this book is domestic urban middle-class life in this country. From this vantage point, some variations of Mundania will be approached. The discussions in the book will start from my own experiences during a specific period of time, the early 2020s. In another time, and in another context, for someone else, Mundania would be experienced differently. I encourage the reader to think about the possible variations of Mundania, and also what are the potential ends of this realm. What is the positionality of different variations of Mundania, its privileges and boundaries? How could variations of Mundania be experienced in different places, by different people, from different positions, in various social contexts and times?

In my discussions I will approach Mundania through a selection of concepts such as media, infrastructures, everyday practices, atmospheres and imaginaries. I will discuss how these different concepts or dimensions could come together in everyday life during the early years of the 2020s. The discussions will be based on stories and reflections about (mostly) domestic everyday life. This takes place physically in Sweden but also, simultaneously, within the clouds, grids and corporate worlds of several (mostly US-based) companies, such as Apple, Google and Microsoft. Domestic everyday life, as it is approached here, seeps through the walls, and stretches beyond that which is defined as a physical home.

During the first years of the second decade of the second Millennium, technological smartness, smoothness and connectivity was hailed by several proponents. It was a time when the future promises and perils of AI (artificial intelligence) were put in the limelight and heavily debated. The development of AI had been going on for some time, but now it became broadly available and transformational. In the debates, questions recurred. Was the new technology friend or foe, saviour or threat? Was it a life transformer or just another tool? Would the novel AI-based technologies be a threat to humans or simply another gradual improvement of technologies? These were some of the concerns and statements. Meanwhile, new practices emerged, regulations and different measures were discussed. All the while, new technological systems and services were launched. Different people used the new powerful technologies. With very dissimilar aims and moral considerations.

The stories and reflections of this book take place right at the time when several new AI-technologies were being broadly implemented. The smartphone was the major entrance point to digital services and worlds, while new devices to engage with digital appearances were introduced. This gives the stories a unique time stamp. It will be a good point of comparison when scrutinizing future emerging technologies. Because new technologies keep on emerging, and Mundania is constantly recreated. The Mundania variations of this book are also characterized by the receding COVID-19 pandemic, by surrounding geopolitical unrest and the erosion of democracies, war, an energy crisis, logistical disruptions and an increasingly unstable climate. A peculiar time of uncertainties and routinization, of novelty and repetition, of the uncanny and the mundane. This is the volatile scene on which I evoke variations of Mundania.

A porcelain insulator

Let us start with a small artefact. A porcelain insulator (Figure 1). It has a shiny glazed surface. It is about twenty centimetres tall and has a curved bell-like shape. For decades it was mounted on a high wooden pole. Metal wire was looped around it. The wire stretched out over the landscape, from pole to pole, from insulator to insulator. Thousands and thousands of factory-produced items. Rows of poles, insulators and wire reached out beyond the horizon.

A certain property of this porcelain insulator is its high resistivity. It prevents electric current from flowing freely. This is the reason all these porcelain items were produced and used. When electric current was flowing through the metal wire, the insulators stopped electricity from spreading beyond the grid. People could walk beneath utility poles, touch them without being scorched by lethal electric currents.

These insulators can still be present in the landscape, but they are also increasingly seldom to be seen. At least that is the case in southern Sweden, the country where this book takes place, the country from which I evoke Mundania. Cables and wires are instead mostly insulated with non-conductive plastic compounds. They are also often buried, and not suspended between utility poles. This is one reason a porcelain insulator stands on my desk,

Figure 1: A porcelain insulator

Source: Robert Willim

no longer employed in the supply of electricity. Instead, I use this artefact, not to insulate, but to spark ideas, to imagine, to probe. The insulator has become a companion in my thinking about Mundania; it is in this sense what Sherry Turkle has called an evocative object (Turkle 2007).

The insulator can be an evocative object or a companion piece when we conceive Mundania and scrutinize ways in which technologies become ordinary in everyday life. When I write this, it is a time characterized by societal and existential uncertainty for many. It is also a time characterized by technology-engendered connectivity, digitalization and transformation. More and more social and societal occurrences are channelled through small screens and speakers. More and more creations are generated or altered by AI-based systems. People interact, engage with and have conversations with incomprehensible technological amalgamations. These occurrences take place through mixes of ethereal flows and heavy industry.

For decades, networks of digitally driven devices and components have been spun together through processes towards an ever more connected world. Uncountable metres of wires have been used to build systems that make devices wireless and unbound. It surely takes a lot of wire to make something wireless. Devices are also supposed to become smarter. A revamp of the old modern mantra of effectivity. This proposed technological smartness can be related to the three words 'connectivity', 'computation' and 'compatibility'. Tech-induced smartness can be evoked by interconnecting more and more advanced computational devices. What will be connected has to be compatible, fitting and aligned with the system.

Imaginaries about rational uniform logic and a pervasive system of systems have characterized much of digitalization. Imaginaries or dreams about pervasive machinic rationality are, of course, nothing new. This has recurrently been associated with technological novelties. Now, in the 2020s, these dreams have arisen with new strength, and they have become incentives to make devices and people compatible and adapted. This has especially happened in relation to the abbreviation AI – Artificial Intelligence or an Alignment Incentive. It has been very much about computation, connectivity and compatibility. What or who should then be made compatible? With what? Wearable devices that should augment and align have been introduced. Extended reality and the instillment of further layers of mediation and demands for compatibility. What was extension here, what was augmentation, distortion (Willim 2013a)?

Here, the derelict porcelain insulator sparks a thought. Increased connectivity also calls for insulation. Everything cannot be connected to everything if people are not to be scorched. Electric systems require insulation, and so do interconnected systems of computational components. How do we approach this circumstance when technological systems are supposed to become ubiquitous, always-on and ambient? This is a question that should be dealt with in Mundania.

In Sweden in the early 2020s, the porcelain insulator was gradually becoming a more and more strange artefact. Many persons born in recent decades had not seen them mounted on poles. And even before then, these items were not tangible or touchable. They were part of the infrastructure. Visible, but out of reach. Visible, but often overlooked by most people. Taken-for-granted, or not even considered, as long as everything worked as expected.

Which novelties were about to emerge in the 2020s? New equivalents to the porcelain insulator. Were they accessible, and something people engaged with? One way to approach these questions is by scrutinizing how new technologies are absorbed into everyday life, and how they gradually become a part in the transformation of the ordinary. A new normal is always emerging. At some times abruptly, through disruptive events, but also slowly, gradually and often unnoticed.

This book is a proposal to imagine differently about the ways technologies become ordinary in different contexts. The book engages with technological circumstances and transformations as a recurring theme, a theme with variations. It appreciates that after a while, most technologies lose their aura. Advanced and sublimely intricate technologies are only experienced as advanced, sublime and thought-provoking for a certain time. Most of the time, they are just there, like the system of poles, wires and insulators. Props for the show. Some tangible, others less so. Once they were porcelain insulators and transistor radios, now data centres and cloud services. When and how are these ignored and enmeshed in the habitual routines of everyday lives?

What is labelled as emerging technologies at one point will gradually escape out of consciousness and debate. Most technologies are not emerging. Technologies might become ordinary or forgotten. Or they can recur, get new meanings, become charged with new energy. Like the insulator. This is a dynamic of taken-for-grantedness, amnesia and recurrence. Variations of Mundania, variations of the ordinary.

The more advanced and interconnected technological systems become, the more it is apparent some questions hover over us: How do the unfathomable complex and even the strange become mundane, and how do new wonders and worries arise to then once again fade away? How is the mundane related to the uncanny? These are some issues related to the dynamics of Mundania, and a derelict porcelain insulator can be our companion when entering this realm.

More than domestication

Mundania is built on mundanization. This word will be our conceptual anchor. I will use it to discuss how the incomprehensible and technologically complex become part of everyday life. While the pandemic had been at its peak at the beginning of the 2020s, homes became places of physical isolation or quarantine. During this period digital connectivity became crucial for sociality and the flow of information in Sweden, at least for those who had the required devices, such as smartphones or computers. The pandemic had accelerated what was called digital transformation in so-called advanced economies (Jaumotte et al 2023). Practices such as video conferencing and online shopping took hold in Sweden on a large scale. Concepts like smart cities, villages and homes had emerged before the pandemic, but now they became even more prominent parts of promotional campaigns and initiatives to install technological systems. A supposed transition should be achieved through networked digital technologies, new gadgets, dense data streams and arcane algorithmic processes of automation, alteration and generation.

How could the sometimes abrupt, sometimes gradual transition be grasped? How does this increase of technology-generated smartness, smoothness, desire and convenience relate to notions of ignorance and control? The increased so-called smartness of ever more interconnected technological systems seems to be related to an increased unknowing among the people that live their lives with and through the systems.

Within media theory, a process through which new technologies become part of everyday life has been framed as a process of domestication (see also Silverstone et al 1992; Berker et al 2006). Most complex technologies are, however, never domesticated. They might be customized and made ordinary. Integrated in domestic settings in various ways, part of routine and the fabric of everyday life. These technologies, however, remain incomprehensible. To a large extent beyond control and beyond grasp. Potentially uncanny, even harmful. They might offer comfort and convenience, a new kind of normality (Shove 2003). They might evoke pleasure and new abilities. They can be transformative and informative, but also deformative and conformative. This is not domestication, not taming. Instead, I suggest the word 'mundanization'. Another way to put it is that the beginning of Mundania is where processes of domestication are supplemented by processes of mundanization. It is an open question where exactly this happens, and how variations of Mundania emerge and evolve. Control and ignorance are not evenly distributed. The fields of unknowing are in themselves a geography that is hard to map or overview. What are people supposed to know? Who and what should be trusted to be in control? Over what?

The emergence of Mundania

The everyday realm Mundania is generated through several processes and phenomena that I will discuss throughout the book. The central process is mundanization, through which technologies become part of the atmospheres, imaginaries and practices of everyday life. Interconnected technologies housing embedded software are integrated in numerous contexts, enmeshed with people's everyday lives.

Technological complexity is veiled and turned into the ordinary, gradually transformed into ignored infrastructures and naturalized patterns of behaviour. Devices connected to other devices and to the internet. Imagery on screens, sound emerging from wireless speakers. Layers that transform perceptions and appearances. Automatically generated creations. Tech-induced conversational partners. It is all there, very present, sometimes pleasurable, sometimes disturbing, oftentimes just prosaically ordinary. Yet, its workings are mostly obscured or ignored. How does this ignorance emerge?

Technological change and novelties often first appear in marketing campaigns and in the enthusiastic presentations by technological evangelists from various corporations and organizations. New words are devised, such as 'augmented', 'extended', 'mixed' and 'virtual reality', 'metaverses' and 'spatial computing'. Coming products and services are evoked as enticing vapourware appearing at an imaginative horizon (Willim 2003a). This is when something can be labelled as emerging technologies, as promises of various futures (Pink 2022). Buzz surrounds it. Attention, curiosity and debate gravitate towards vapourware and emerging technologies. Then, some vapourware turns to concrete offerings, often launched through events and campaigns.

When introduced and at hand, the new technologies can appear as utterly fascinating. Some new possibilities can be awe-inspiring. The new can also be experienced as awkward, even disquieting. Irritating, or even threatening. Sometimes novelties can, despite the loudness of marketing campaigns and debates, be experienced merely as small iterations of earlier technologies, services and products. If successful, all this, the fascinating and the awkward, soon disappear in the muddle of everyday life. As Wendy H.K. Chun has noted, 'our media matter most when they seem not to matter at all' (Chun 2016: 1). When devices and systems become habitual, experienced as natural parts of everyday life, then they really start to make a difference. This is when all the buzz around the technologies turns to a faint hum. This is also often when technologies might slip out of control. They become ever more integrated in people's lives, while their workings are oftentimes ignored. This is how Mundania takes form.

During the first decades of the 21st century, internet, social media, cloud services, and advanced algorithmic operations had step by step become part of the variations of Mundania. New layers of technology keep on persistently becoming parts of habits and everyday environments. The intricacy of the underpinnings of it all increases as new layers are added. The ground and the atmosphere shifts, while life can be experienced as continuously ordinary. When new technologies and systems are mundanized and when they work as expected, they are often experienced as elementary parts of daily practices part of standardized arrangements (Bowker and Star 1999).

Think about how ubiquitous computing was advocated by Mark Weiser and his colleagues at Xerox PARC (Palo Alto Research Center) in the 1990s when they envisioned future technologies. Ubiquitous computing, or pervasive computing, was promoted as becoming ambient, part of the atmosphere of everyday life. They also talked about this ambient technology as calm technology. 'The most profound technologies are those that disappear. They weave themselves into the fabric of everyday life until they are indistinguishable from it', as Weiser formulated it in the early 1990s (Weiser 1991: 94). Computing as calm, ubiquitous and ambient. Decades later some talked about computing and AI as part of an emerging ambient intelligence (Gams et al 2019). New layers of pervasive, and grid-based technologies were about to become ambient. These systems have no obvious on/off button. Instead, they become entangled with the ambience or atmosphere of everyday life.

What we see here is a drift between attention and disregard, between embodied skills and ignorance. When internet became broadly available during the 1990s, around the time Weiser proposed calm technologies, it was noticed differently than some decades later. To connect in a home in Sweden in 1995, a set of concrete actions were required. People had to connect to the internet using a sometimes cumbersome procedure, and had to know how to perform this procedure using paraphernalia such as modems. People called up/connected to the internet, and then after some time disconnected. Some decades later, internet was more like a basic utility, a constant supply of ambient connectivity. The intentional action and procedure of 'calling it up' or 'going on the web' disappeared from everyday life when internet became more or less part of the atmosphere.

In 2006 Adam Greenfield captured how computing technology moved beyond the personal computer by using the word 'Everyware'. Since then, the discourse has shifted more towards the Internet of Things (IoT) and Smart Technologies and how they relate to algorithmic operations, AI and platform economies (Larsson and Andersson Schwarz 2018; Sadowski 2020). We need to know more about these processes, about the ways through which technology becomes or does not become ordinary. How and when does the novel seep into people's lives? Infrastructures, supply-chains, and distributed power relations are rarely thought upon when technologies have been effectively integrated in Mundania.

In 1991 Weiser wrote about technologies that became part of the fabric of everyday life as calm technologies. Almost thirty years later, some technology critics proposed that this calmness had turned into sadness (Lovink 2019), as part of a colonization of everyday life (Greenfield 2017). According to these critics and scholars, a quite dark emotional tenure permeates the uses of social media and the latest internet services. According to internet critic Geert Lovink, these technologies are *Sad by Design* (2019), causing negative affective feedback loops. Several arguments against too optimistic ideas about digital technologies have been proposed: warning that techno-utopianism is dead and sounding the alarm that 'Your Computer is on Fire' (Mullaney et al 2021), and suggesting that we live in a highly troublesome time of 'surveillance capitalism' (Zuboff 2019). A New Dark Age (Bridle 2018). A time of systematic tech-induced inequality (Broussard 2023).

Calm or sad? On fire or comfortably warmed up? Repressing or exciting? Maybe these affective varieties coexist in Mundania, in the atmospheres of everyday life with complex technologies? Feelings and affective formations are often fleeting and ambiguous (Paasonen 2021: 11-12). Mundania is an ambiguous realm, based on tensions and paradoxes.

Mundane and uncanny

Much technology seeps into everyday life without being heavily scrutinized, and it becomes part of the transformation of lives. A recurring phenomenon is furthermore that at some point in the life cycle of a popular technology, critique and discussion seems to vanish, sometimes but not always to re-emerge. Even though there are obvious uncertainties and even risks, many technologies are still used. People live in and with systems that they suspect are unsound and unfair, but what to do about it? Worries and discomforting feelings are subdued and reside in the periphery like a faint atmospheric hum.

The early 2020s was a time when several technologies emerged. Some were supposed to be intelligent. The word 'smart' or 'smartness' was used to promote new technological development in domestic settings (smart homes), or as part of ideas about sustainable and 'future-friendly' places such as smart cities and smart villages. The use of this word to refer to advanced technologies had been going on for some time, and its uses had also been scrutinized (Heckman 2008; Strengers 2013; Parikka 2017; Kitchin and Dodge 2019). What could smartness imply in Mundania? Smart could be related to smooth. Complex things that through smart designs could be experienced as easy, as smooth. Here, the smart could be seen as part of a solution. The question is: What should be solved, and what remained outside the supposed instrumental logic of a recognized problem that needed a specific solution? What was ignored or even concealed in different offers? While several technologies, services and what in marketing campaigns were called 'solutions' were promoted and often adopted, they also paradoxically introduced new (and reproduced old) problems and risks (Morozov 2013; Kitchin and Dodge 2019). This is not domestication, it is mundanization. It can also be related to how the normalization-concept has been used in studies of power and inequality. Within gender studies normalization is used to understand how obviously malevolent patterns such as recurring violence can also be experienced as normal in relationships and human organization (Nilsson 2013).

This is how the mundane can also be creepy, uncanny. What could in some circumstances be experienced as frightening is instead subdued, felt and considered to be mundane. This mundanity, however, comes with an undercurrent. As Sherry Turkle puts it: 'The uncanny is not what is most frightening and strange. It is what seems close, but "off," distorted enough to be creepy' (Turkle 2007: 8). When technologies are mundanized and things appear as usual in Mundania, the creepiness is mostly subdued. Like a faint hum that mostly goes unnoticed. Wendy Chun has proposed that new media are wonderfully creepy. That they 'are endlessly fascinating, yet boring, addictive yet revolting, banal yet revolutionary' (2016: ix). The tensions between the mundane and the uncanny, the ordinary and creepy, are the ambiguous qualities of Mundania. Tech-infused life in Mundania can be comfortable, convenient and uncomplicated, but sometimes its uncanniness, its creepiness can become apparent. Once in a while, people might notice it. When something glitches, or something unexpected happens (see also

Krapp 2011). Or when someone suddenly seizes and begins to reflect. Those mundane epiphanies that sometimes occur. What are those fixtures, really? What is this networked glass and aluminium device that I hold in my hand? Handheld, yet interconnected and ungraspable. That automatically generated stuff that appears on its screen? The alterations. A constant flow of appearances. Where does it come from? A soothing voice coming from a speaker, filling the room. I respond to it. It responds to me. Who, or rather what, do I have this conversation with? It is all so familiar, yet so incomprehensible. In the atmosphere, yet so distant. So clearly outlined, making sense. So present, yet ungraspable. Mundane, uncanny, convenient, creepy, ordinary and weird.

Despite the subdued creepiness and weirdness, mundanized devices, systems, technologies and circumstances are almost impossible to bring up for discussion or in small talk. Technologies that once were emerging, but have become integrated in everyday life are more atmospheric than discursive. Part of an atmosphere of unquestioned but also sometimes reluctant acceptance. We need to know more about when and how mundanization really happens. This transformation. Despite widespread awareness about potential threats, disturbing circumstances and injustice, attentiveness often turns to ignorance when emerging technologies are mundanized.

Mundania is to a high extent about unknowing, and about, furthermore, often unequal distributions of ignorance and control. To trust the implementation of several new technologies, people were and are often supposed to trust what is going on beyond their understanding. In this sense, the early 2020s was a time when the smartness of technologies and the smoothness of experiences was often promoted and bundled with comfortable ignorance (Sadowski 2020b).

Variations of Mundania

Then, what stays in Mundania? What makes a technology widely consumed, and for how long? It depends on schemes, deals, arrangements, and competition between various stakeholders. There are, however, other issues that determine what becomes successful or not. New technologies are not simply poured over a population that just swallows everything. Incorporation of technologies in everyday life differs between contexts, and it doesn't always happen in a smooth way. Some technologies are never accepted. There is resistance, friction and debates, as well as controversies on issues such as integrity, autonomy, power and control. Regulations are sometimes introduced. Often, they come first when some obviously serious issues have occurred. There is also tacit and tactical resistance going on (Velkova and Kaun 2021). Several technologies and services are discontinued and never reach any larger success. Some remain as fringe occurrences. Others are, however, in various ways becoming enmeshed in people's lives, and sometimes in ways not intended by developers, promoters and planners. 'The street finds its own uses for things', as science fiction author William Gibson once put it (1982: 106).

Processes of mundanization and the ways in which Mundania is engendered is context-dependent and varies over time. Mundanization is an embodied as well as social process that appears when technology is consumed and integrated in everyday life. This is how different variations of Mundania emerge. Mundanization will be different in various temporal, spatial and social settings, and it has its rhythms, choreographies, and tempi (see also Jalas and Rinkinen 2016; Blue 2019; Spurling 2021).

Even if it might be problematic to fully embrace thoughts about a specific zeitgeist or the spirit of the times (Warde 2021), it is important to still take the specificities of broader temporalities into consideration, even if these might be patchy, ambiguous and somewhat contradictory (Paasonen 2021; see also Champion 2019). The 2020s differ from the 2000s, but there are also processes that remain and have been going on for a longer time. It can be valuable to pinpoint some of these differences and continuities. To do that, an analysis of the kind I propose must work across different scales (Eriksen 2016). It must also bring up examples of how earlier technologies were introduced. What can the 20th-century introduction of radio or electricity in domestic settings say about the way Mundania is experienced, imagined and practised today (Löfgren 2021)? Small-scale mundane practices and whereabouts should be seen in relation to broader processes and different temporal conditions. It is, however, important to note that my proposal to see everyday life as life in Mundania is not based on any attempt to write a history of media, technology and everyday life. Neither is it to make detailed empirical accounts of different contexts.

Once upon a time in the West

Processes of mundanization are, as noted, context dependent. When telling stories and evoking thoughts about Mundania and its variations, we must, however, start somewhere. Sweden in the early 2020s might be a good point to evoke Mundania, to better understand and imagine what mundanization is and could be.

Sweden has a long history of embracing technological progress, as well as engineering, on a policy level. Digital technologies have been profoundly encompassed by authorities, businesses and by large parts of the public. When internet infrastructures and new ventures based on digital technologies were developed around and before the turn of the Millennium, Sweden was understood to be at the forefront of this process (Willim 2003a and b). In the coming decades, the Swedish government would stress the importance of digitalization. It became an official aim that Sweden should be best in the world when it comes to digitalization and to harness the potential of new digital technologies (*Regeringens skrivelse* 2017/18:47). The vision was that the whole country should become connected, and digitalized in a sustainable way (*Regeringens skrivelse* 2017/18:4).

What was expected as the normal in early 2020s Sweden was to use digital services, to have a digital ID (or e-identification), email and a smartphone or similar device, to be online, to be connected. Digital technologies and interconnected systems of infrastructures and techno-organizational assemblages were increasingly enmeshed in people's everyday lives. With variations, of course. Taken-for-granted, but in different ways (Ling 2012).

Despite the heavy focus on new digital technologies in Sweden, they were not available to all, not part of everyone's everyday life. I write from my particular position, from the middle class in a Western country. From an urban setting in southern Sweden. Questions about outright inequality are not in the foreground in the stories or analyses. Neither are different circumstances pertaining to race, gender or abilities. Nor poorly working institutions or infrastructures, extensive societal disturbances, or oppression, which is part of people's lives in many places around the planet.

Inequality and disquieting practices are, however, still entrenched in the variations of Mundania I evoke. It is entrenched but also subdued. This entrenchment is what makes the habitual use of extremely advanced tech not just smooth but also uncanny, creepy, weird. Where do the pristine devices that I unbox come from? Whose hands have touched the components and material, and under what circumstances (Chan et al 2013)? Where do devices go when they are discarded? What about all the unspecified harvesting and massaging of personal data that is inscribed in various kinds of deals and arrangements? How are relations and selves moulded by the interaction and engagement with more and more incomprehensible assemblages of AI and corporate operations? Or all the outright wicked violent and cruel uses of technologies, sometimes performative, sometimes covert. Or poor working conditions, even exploitation, harmful handling of e-waste or personal data. It is not clearly present in the variations of Mundania I evoke or propose. Or rather, it is there as that faint hum, an uncanny notion, a dark tenor and an often disregarded suspicion. Part of the fabric of ignorance that veils Mundania.

How then could inequality and other malevolent circumstances be disentangled? Through perfectly working technologically engendered order? Through another layer of technological solutions? No. Inequality, and even oppression, can instead be maintained and enforced by too strong a focus on technocentric solutions. Meredith Broussard has called it 'technochauvinism'. This is a phenomenon fuelled by imaginaries about machinic order, justice and the dream that rational, ever more advanced, technological systems can generate a better society (2023: 8). The complexities of the social cannot merely be solved by computational or mathematical logic. Therefore, technological problems are often 'more than a glitch', they can be more fundamental (2023: 8). Inequality and other complex social challenges have to be dealt with through non-tech-centric approaches that take ambiguities and humane, as well as more-than-human, dimensions into consideration.

Abilities and variations

Technologies can give new possibilities. They can also come with challenges. Several technologies and systems might be intrusive, and they might increase maladies such as inequality. Architectures, designs and technologies also influence what is experienced as ability or disability. What is required from a human who is supposed to live with certain technologies? Abilities? Adaptation? Alignment? Everything from minute details such as the size and colour of signs and other features of devices to the architecture and planning of systems might suppose that people have certain abilities. Certain designs can even be 'disabling barriers' (Egard and Hansson 2021). Design can induce context-specific disability. The positioning of buttons or panels can make them impossible to reach for some; the size and shape of text and symbols could require more or less perfect vision; digital services could be absurdly complicated to comprehend, navigate or interact with.

The specific appearance and design of a technology might be something that hinders some people from participating and being included in a just way. But technologies relate to abilities in multiple ways. Some technologies are also life-supporting and life-enhancing aids. People with disabilities might enrich their lives through digitally engendered means. People might, for example, find meaning and abilities in virtual worlds, as anthropologist Tom Boellstorff has shown (2015). Technologies can challenge norms. What is considered as normal sensory perception such as hearing or seeing can be questioned when using technological enhancements (Friedner and Helmreich 2012). These different examples hint at the ambiguities of human life with technologies. Technologies can be both remedy and malady. We have to keep on analysing when and how this happen.

In my analyses I will hint at the mundane, and often banal, but also at the uncanny, and at sombre dimensions. However, as already noted, from where I evoke Mundania, the uncanny and the sombre is mostly noticed as a faint background hum, not as something that hits straight in the face. The question is when the processes of Mundania shapeshift and turn into something outright hostile. The question is also how Mundania would emerge and be manifested for different people, under different circumstances. In this book, I will discuss and juxtapose several quite different examples and theoretical points of departure to evoke thoughts and imaginaries, foremost from my perspective and experiences. But Mundania could emerge in different ways. It is a proposed realm, but it is not a clearly spatial concept. I play with its spatial whereabouts, but it is a more fuzzy and imaginative evocation. A theme with variations.

Why this book?

What is the reason to write yet another book about technologies and how they relate to people's lives? Although emerging technologies, media, digital cultures, and everyday life are heavily studied fields, we still need new approaches to understand how the unexplainably complex become part of everyday lives. The fantastic and the upsetting, the troubling and thrilling, is recurrently turned into the mundane. How do emerging technologies turn from vapourware, via novelties, to invisible and ignored arrangements? Why are some technologies successful and others not, and what does success really mean here? Malevolent technologies and practices also become part of the obfuscated fabric of everyday life. How is the uncanny tightly interwoven with the mundane and even the banal? What are the different dimensions of the ordinary? Ordinary technologies? How, when, why and for whom?

By evoking Mundania, I attempt to mash up some of the discussions and presumptions about emerging technologies. We are inevitably bound to the present, with its concerns, circumstances and debates. When trying to analyse technologies and everyday life and to imagine futures, we need to experiment with different methods, to destabilize all the present excitements and concerns. I have chosen to not concentrate on the specificities of the most hyped and debated technologies that appeared as this book was being written. In the timespan between writing this and the time it is read, several novel technologies will be released, some successful, some not. Everyday lives will house different devices, practices and protocols.

It is easy to become confused by the onslaught of what is just emerging. Therefore, I do not concentrate on what headlines the technology debates, even if much of the technological landscape is about to transform. Instead of focusing on the points of most turmoil, I explore some of the fringes of what is happening in the early 2020s, and I also examine that which is already taken for granted. I concentrate mostly on fringe devices, and the supposedly unexciting or that which is just about to become ordinary. Some key insights about emerging futures might be found in the bland and ignored, as well as among peripheral occurrences. My tactic is an attempt to avoid the noise surrounding what is most hyped, debated and promoted. Since this noise often dichotomizes and seems to push analyses and thoughts into hyperbole and unproductively simplified stances for or against certain technologies.

My belief is that we need imaginative approaches driven by curiosity more than predefined agendas and fixed positions. We do not only need new knowledge about complex circumstances, nor only swift action. We also need to imagine differently, and hopefully thereby open the locked positions that often recur in debates about technologies. By proposing that emerging technologies recurrently withdraw as part of a process of mundanization, and that this induces the everyday realm of Mundania, curiosity and the wayward might be put at the forefront. New thoughts can hopefully be provoked in debates and in academic fields already characterized by an abundance of theory and studies.

My background

Since the 1990s I've done research around tech-related phenomena and tried to understand the role new digital technologies have in people's everyday life. How do technologies transform everyday life, and how does everyday life transform technologies (Löfgren 2015)? How do imaginaries relate to practices and how is life entwined with infrastructural, technological, corporate and organizational processes? I have for a long time been interested how the concretely material is related to imaginaries and how mundane practices unfold. The porcelain insulator that introduced the book is a good example of how I use artefacts to spark analyses, stories and imaginaries. By juxtaposing sometimes very different things, concepts and phenomena I let stories unfold.

My intention is to extend discussions that have been going on in media studies, in art and design, and in social and cultural studies of technology for some time. I will touch upon several academic fields in a quite eclectic way. I will relate to design-oriented research within usability studies and Human Computer Interaction, as well as information and computer science (Norman 1998; Dourish 2016) and how this research can also be associated with advances in, for example, design anthropology or design ethnography (Gunn et al 2013; Pink et al 2022). The book is inspired by phenomenological and arts-based approaches (Willim 2017b; 2023). My arguments and examples will be related to discussions within digital cultures research and involve accounts of the role of imaginaries (Willim 2017a; Mager and Katzenbach 2021) and aesthetics to mention some aspects.

In my research I have studied how the first internet consultants in Sweden around the shift of the Millennium promoted and contributed to the rise of a society permeated by digital technologies in which everyday life is dependent on uninterrupted internet connections and arcane techno-organizational workings (Willim 2003a, 2003b). At this time the internet was often associated with imaginaries about *Cyberspace* and what was called *The New Economy* (Löfgren and Willim 2005). A time when new technologies became part of new economic practices and imaginaries, when North American frontier imaginaries and the pull of Silicon Valley was strong (Barlow 1996;

Kelly 1999). Google was not yet part of most people's lives, Facebook was not yet founded. Netscape, with its pervasive web browser Navigator, was a highly influential company for some years until Microsoft took over with their browser Explorer. Then, after some years, Google grew strong and a period of extensive search engine practices started. The corporate landscape of Mundania changes over time, and what for some time is experienced as a persistent business and market logic can be quite ephemeral.

As an extension of my research about internet and everyday life I have also studied imaginaries about factories and industries in societies that have been promoted as being postindustrial (Willim 2005a). In this research I analysed how the industrially infrastructural was highlighted, showcased and imagined in some novel and not so novel ways, foremost in Sweden and Germany. This was a good way to start scrutinizing the dynamics of technological visibility and obfuscation, and how this played out in relation to ideas about transparency (Willim 2005b).

My academic background is in ethnological cultural analysis as the discipline was developed at Lund University in Sweden since the 1990s and onwards. The core of these practices has been a heterogeneous approach, using a mix of methods, theories, and material. Interplay between the serendipitous and the systematic has been a recurring characteristic of this strand of research (Ehn and Löfgren 2010: 217-218; see also Ehn et al 2015).

Probing

In my work I entwine research with artistic practice through an open-ended process I call *probing* or *art probing* (Willim 2017b; 2023). As part of this process, I work with different methods and formats, everything from the writing of texts to the production of music, sound art, performances, installations, sculptures and experimental film. The porcelain insulator has recurred in many of my art projects.

The way I use the words 'probing' and 'probes' can to some extent be compared to probes as scientific instruments of natural science. But in a metaphorical sense. The way I use the word is more related to the verbal probes used by Marshall McLuhan (McLuhan and Carson 2003). For McLuhan, probes were short statements intended to evoke associations and provoke thought experiments. His probes were often ambiguous and puzzling, they were phrases meant to tease or even disturb. These probes were also used as part of his explorations. Some of the most well-known are the wordplays between message/massage in 'The media is the massage' or phrases like 'The most human thing about us is our technology' (McLuhan and Carson 2003).

When referring to McLuhan, I am fully aware that much of his writings are questionable and problematic. Throughout his career McLuhan proposed a swarm of ideas about humanity, media and technologies. Several were provocatively trickster-like exclamations, but many of his thoughts were astute. Media scholar John Durham Peters, who has written about these matters, also points out this catch: 'Much is maddening about McLuhan—his obscurity, mischievousness, and willingness to make up or ignore evidence— but his brilliance covers a multitude of sins.' (Peters 2015: 15). A similar stance as Peters' is taken by Sarah Sharma and Rianka Singh et al (2022). They extend and tweak ideas from McLuhan and put them in a new and feminist critical context.

Words and concepts I suggest as part of my discussions throughout the book can be seen as probes. I launch them and see how far they fly, what they do and how they work. In my doctoral dissertation and book about Swedish internet consultancies around the shift of the Millennium, I have already used this method by suggesting cultural analytic concepts such as *The Rosebud-syndrome* and *conceptual congruity* (the latter will appear later in this book) (Willim 2002, 2003a, 2003b).

Probing, as I understand it, is, however, not just a play with words and concepts or about evocative aphorisms. Probes go beyond the linguistic. Through art projects I have explored concepts from my research such as Industrial Cool, Possible Worlds and Mundania in a practice that extends beyond the linguistic and the academic. To some extent this way of working with probes has similarities with ways in which design probes or cultural probes have been developed (Gaver et al 2004). But I mostly use probes beyond specific empirical or design-oriented contexts. I work with sound and music performances, with video art and experimental film to open-endedly explore imaginaries, materials and atmospheres related to the concepts I also use in my research (Willim 2017b, 2023). Probes travel between art and research and are ways to spark ideas, to probe how conditions might or could be, how they become or could be transformed. Probes provoke questions. They also work affectively, and they challenge the way we deal with the empirical as well as the theoretical (see also Dunne and Raby 2013; Ssorin-Chaikov 2013; Vannini 2015; Moores 2021).

The methodological stance I advocate breaks with ideas that the practices of research should follow a linear and preconceived trajectory. Here, my position resonates with that of several other scholars. A predictable linearity of research, delineated data sets, and the possibilities to design research projects beforehand has been broadly questioned, challenged and discussed, especially within different parts of qualitative research. In these strands of research, what takes place is often an interplay between the methodological and the irregular and serendipitous (O'Dell and Willim 2011a; Wilk 2011). Researchers engage with the world, with devices, places, and relations, in ways that influence research in often unforeseen ways (Löfgren 2014: 116; see also Lury and Wakeford 2014; Law and Ruppert 2013; Löfgren 2015b).

Qualitative research is, of course, planned, methodological and structured, but it also takes form as researchers move along more open-ended processes.

To challenge rigid forms of academic research, anthropologist Tim Ingold have developed what he calls an *art of inquiry* (Ingold 2018). He acknowledges that much research has similarities with the creative practices of making, within, for example, art. Another strand within social sciences and humanities that can be associated with both Ingold's approach as well as my probing practices has been developed within cultural, or human, geography and sociology during the last decades. This strand has been called non- (or more-than-) representational theory and methodology, and has been advocated by scholars such as cultural geographers Nigel Thrift (2007) and Hayden Lorimer (2005), media scholar Shaun Moores (2021) as well as sociologist Philip Vannini (2015), among others. This methodological approach is often amalgamated with creative practices and art (Boyd and Edwardes 2019).

Another way to reframe qualitative inquiry that corresponds with my practices of probing is what Annette Markham has called a remix approach, or remix methods. It highlights 'serendipity, playing with different perspectives, generating partial renderings, moving through multiple variations, borrowing from disparate and perhaps disjunctive concepts, and so forth' (Markham 2013: 65).

Remix methods also resonate with the way Ehn and Löfgren have worked with bricolage and an eclectic cultural analysis (Ehn and Löfgren 2010; see also O'Dell and Willim 2011b, 2013). These approaches have been framed as ethnographic. The way I work with probing is based on artistic explorations, but it also has autobiographical or autoethnographic features. During the last decade the ends and beginnings of ethnography have been heavily debated, and ethnography has become a widely used word (Ingold 2014; O'Dell and Willim 2017; Willim 2017b; Rees 2018). Several scholars have also taken the second half of the word 'ethnography' and promoted new practices such as praxiography (Mol 2002), technography (Kien 2008), digital technography (Berg 2022) or autotechnography (Hildebrand 2020). Here, I will instead lean on a combination of art and cultural analysis as a practice of probing.

Range

The work with this book has been based on engagement with digital, but also non-digital, technologies, foremost in domestic settings. I have used my own experiences as a point of departure; I have observed and engaged. I have taken part in different events, such as openings of infrastructure facilities, debates and marketing events. I have also made observations in shops selling electronics or furniture, as well as at a large furniture fair in Cologne, Germany (IMM Cologne 2020 – The International Furnishing Trade Fair), which took place just weeks before pandemic lockdown. I have had conversations with people about connected homes and digital cultures, and I have studied numerous stories and discussions about technology and everyday life in news media, in debates, in documentaries and on various sites and feeds, from X, which was then called Twitter to YouTube.

These examples of methodological approaches are what could be considered ethnological or cultural analytical ways to proceed. I have used them as resources for a more extended process of probing. What has become especially foregrounded in my work is a constant movement between cultural analysis and processes of making and engagement with artefacts and technologies. Mixes of art and cultural analysis based on explorative practices of making and the arrangement of material, and employment of techniques and technologies, are central to the way I understand probing. I have conducted art projects and engaged with technologies and related this work to arguments and discussions in my research. This probing has been my way to approach technology-permeated everyday life in Sweden and to imagine this as the emergence of the everyday realm Mundania.

In some of my probing projects, I have explicitly taken the domestic and the mundane as the point of departure. In these projects, the challenge and motivation has been to start with the most banal circumstances to tell stories and evoke imaginaries. In this book I also start with prosaic situations, many of them in my own home, that have spurred stories. They are journeys round a domestic geography, travel narratives through the magnificently mundane landscapes of press buttons, screens, floorboards and cables (cf. de Maistre 1871; Willim 2013b). This home-based geography is anchored in a specific social setting that could be labelled Swedish urban middle class. It is a geography permeated with specific services and products, from specific vendors and producers. In another domestic setting, the products and services could come from other stakeholders. The social and economic circumstances could be different. Mundania could be experienced differently by different people.

This is the methodological background to the upcoming discussions and stories. Probing can be conducted in many ways. Mundania can also be evoked in several ways. Here, it is in the shape of a book. My aim with this volume and in evoking Mundania is not to build an edifice of philosophical rigour or appropriateness as envisioned from within some specific academic discipline. Neither is it to make a thorough empirical account of emerging technologies or domestic mediascapes of the 2020s. Instead, it is to propose an imaginative exposition that can be used when analysing how technologies become ordinary in everyday life. The book has an essay-like style that conveys stories that intensify the ideas about Mundania. As a part of my probing practices, I juxtapose several quite different examples to provoke thoughts and associations. Shorter theoretical expositions are combined with arguments as well as stories resembling diary entries. When I describe experiences of technologies and services, it is from what could be called a user perspective, or rather as part of everyday life where the end and beginning of usage is not clear-cut. Through this mix I want to spark thoughts about a shifting everyday realm where emerging technologies withdraw and are gradually ignored.

When I suggest that Mundania is a proposal to imagine differently, it is also to some extent a theoretical construct. Throughout the book, I will discuss and refer to different scholarly approaches. However, I will approach theory as something that is always in the making, more like a poetic device than a finalized statute. Something narrated and composed, with rhythms, repetitions, juxtapositions, references and rhymes. I consider cultural and social theory as enacted, time-based and context-dependent.

I do not use theory in a confessional manner. I do not agree with everything certain scholars and intellectuals have written, but I might still use some of their insights and writings if it is relevant for the present work. My use of McLuhan is a case in point. Therefore, I will not spend too much time denouncing theories or arguing against them. Neither will I bring up and position scholars and thinkers against each other. My approach is more about composition than polemics. Academic practice, as I aim to conduct it here, is a shifting process, not the establishment of a set of static statutes. It is based on careful choice and examination, but also on openness for the unexpected and the serendipitous.

This means that the configuration of theories and stories I assemble to evoke Mundania is merely a temporary equilibrium. A theme with variations. Maybe this theme will be relevant and repeated for a while. Used, applied, developed, revised or remixed. For a month, a year, a decade? Time will tell. If used, then also transformed. If not used or repeated – forgotten. These are the possible prospects of Mundania.

Outline

The structure of the book is based on eight substantive chapters, in addition to this introduction, and I will present and discuss concepts and several theoretical arguments and tell stories that may be relevant for the exposition of Mundania. Step by step, a suggestion of how technologies can be made ordinary within Mundania will be sketched out.

Mundanization is about relations between the incomprehensible and the concrete. Between the ephemeral and hands-on everyday life. I will propose ways to approach these relations. I will interweave concrete examples with theoretical discussions and juxtapose different scales, wide-ranging processes and minute details.

Chapter 2, 'Vanishing Points', deals with concrete engagement with devices. I discuss some very small details of domestic technologies to evoke how Mundania can be built from the tiniest of the tiny. Small in appearance, but often great in impact. These are points of engagement, where people meet technologies. But they are also vanishing points. At the end of the chapter, I will suggest three orientations, to suggest how technologies of everyday life withdraw and escape attention, how they seemingly vanish. These orientations will be dealt with in the three following chapters.

In Chapter 3, 'In-Between' I discuss the atmospheric and the in-betweenness of technologies that turn ordinary. It is about environmental media, and about what tends to ethereally dissolve. In Chapter 4, 'Beyond' I discuss how *elsewhereness* seeps in and out of everyday life with complex technological arrangements (see also Willim 2013b). The chapter deals with imaginaries and geographical distance. In Chapter 5, 'Beneath' I discuss the infrastructural in relation to Mundania. It brings up that which is infrastructurally concealed and that which normally stays below the threshold of attention.

In the following three chapters, various properties of mundanization and Mundania are suggested. In Chapter 6, 'Opacity' the role of concealment is the point of departure. Here, the cryptic and the arcane, as well as transparency, are discussed. In Chapter 7, titled 'Order', come discussions about arrangements, standards and routines. It also deals with the ambiguities and intricacies of everyday life with technologies. It discusses the tension between the orderly and the messy and how this tension characterizes the way Mundania keeps on gradually emerging. This leads to Chapter 8, in which 'Variability' and change is brought up. It discusses how the ordinary relates to transformation. It is about different experiences of change, interferences and suspensions, and it further points out that there is variability in Mundania. It also poses questions about the possibilities to imagine differently.

The chapters consist of a heterogeneous theoretical theme that is gradually developed. It is interspersed with examples and stories. These are small variations on the Mundania theme, and they all start with some concrete situation or occurrence that differ in scale and character. It can be about close engagements with small details on a screen, or about experiences of the nocturnal firmament, and several phenomena in between. The variations are diary-like and time-stamped and will introduce lines of thought in a quite meandering way. The time stamp acknowledges that technological, as well as other circumstances are shifting. In this sense, the examples or variations say something about conditions during certain specific years. In a specific context. There lies an analytical potential in comparing these circumstances with the way life with technologies can be experienced and understood in the time and context in which the book is read. What is different, what has changed? The variations also act as counterpoints that will contribute to multiple understandings of the Mundania concept. The theme and variations will together be oriented towards the aim of the book: to evoke thoughts and imaginaries about the ways in which technologies can be made ordinary in everyday life.

Lastly in the book, there is no closure. Instead comes Chapter 9, 'Openings'. A kind of provisional terminus or portal and some suggestions to move on. It further extends the variations and tensions of Mundania and presents the role of dilemmas and questions. It shows and deals with the fissures of Mundania and tries also to discuss how there might be openings in the construct. The chapter is formed as a provisional cabinet of curiosities. It exhibits small stories that are openings and suggestions for the reader to formulate similar stories and to ask new questions. How can the strangeness and the peculiarities of Mundania energize, how can they become part of an open-ended exposition and be vehicles for the imagination?

2

Vanishing Points

Devices, gadgets and everyday things might be the most obvious tangible entities of Mundania. Present technology. Perceivable and touchable. But what does it mean that something is tangible, and for whom and when is it tangible or graspable? How can different aspects of technologies be controlled in Mundania, and by whom? What do different people have to know about the technologies they live with? What can they know? The aim of this chapter is to approach these questions by closely engaging with some of the devices, materials and details of Mundania, and to start thinking about how they often seem to vanish.

Hands-on

To grasp something is to gain control, to have it at hand. To notice, comprehend and get in touch with it. To engage with processes, relations and devices (Dahlgren and Hill 2022). There are power dynamics at play here. Who can grasp what, who or what is under influence, and what do different people come to grips with? What does it mean that something is hands-on in a society permeated by complex technologies?

Many societies of the early 2020s are built on vast interconnected technological and organizational systems. Large-scale systems are furthermore built of miniscule components that are impossible to notice or adjust without advanced instruments. Top-secret facilities like data centres and connection points, housing a plethora of interconnected infinitesimal components.

The technologies of Mundania are ungraspable due to scale. Too large or too small. They are also ungraspable due to spatial or conceptual perimeter control. Many sites and whereabouts are secret, shielded and fenced off. The assemblages of technologies are huge and microscopic, distant but also close. When close and at hand, often also boxed in, sheltered and shielded. 'Warranty void if seal is broken.'

Even though much of the world of advanced digital technologies is out of grasp, the word 'digital' pertains to something utterly concrete and human, namely fingers. Digital comes from *digitus*, the Latin word for finger. Throughout history, the fingers of a human have often been used for counting, and early calculating devices like abacuses are used with fingers. Devices for calculation operated by fingers are early and graspable forebears of today's systems of computation.

While more and more devices and systems become based on voice control, finger-based operation is still a central way of dealing with present digital devices. Taps on keyboards, gestures on touchscreens, trackpads and the physical handling of devices. There are many designs and constructions that have been developed between the abacus and the smartphone, constructions involving a plethora of components and features. One such basic component is the press button. Our world is permeated with these. A press button or a switch often has a diminutive simplicity in its physical design. But what can a button do? The small physical appearance of a button does not immediately reveal all the workings and manoeuvres it can evoke. Buttons are minor, but often significant.

In a history of the power button, Rachel Plotnick shows how press buttons were far from any mundane phenomenon when they started to be broadly implemented some centuries ago (2018). A press button could be and can be highly controversial. Someone presses a button and initiates a process. Or someone flicks a switch, and something happens. This minor bodily action can be an action of command and control. Systems based on certain kinds of minor manual actions could facilitate much larger effects than the small bodily move would reveal.

Plotnick relates button pressing to a fantasy of what she calls *digital command*, through which '[certain] hands could direct anyone or anything to submit to their will' (2018: 227). A button can be an instrument of power. All, of course, depending on what the button is connected to. It can start machineries or processes; notify, open or close down; activate or terminate. Plotnick writes about the politics of pushing. How influence and authority could be mediated by the fingertip. The more advanced the systems that utilized buttons became, the further the reach and consequences of digital command stretched. Button pushing spurred action across distance, and the effects generated could be beyond what a button pusher perceived and experienced directly via eyes, ears or the rest of the body. As button pushing moved closer to people, and became part of everyday lives, the reach of actions extended further (2018: 228-229).

The fantasy or the imaginaries about digital command spurred ideas about armchair commanders. Workplaces and domestic settings were soon filled with buttons and switches that facilitated a plethora of actions at a distance. When an electric light was switched on, a stretched-out electric system was employed, engaged and consumed. This included transformers, wirings, electric insulators and distant powerplants. Somewhere in the landscape of electric distribution there might have been porcelain insulators. Mounted up on poles. Visible, but unreachable. When pressing a power button, a person was also engaging with the insulators in the landscape. Buttons became means to also engage people at a distance. Sometimes indirectly, sometimes directly. By pressing a telegraph key, messages could be sent to other parts of the world. This message across distance could spur action, set stuff and people in motion. The telegraph key was an early remote control, connected to a vast system, a Victorian internet (Standage 1999).

Today many homes are speckled with press buttons, small devices connected to larger systems. A good question is who use to press the different buttons in a certain space. Who are the different button pressers in various contexts and who were the button pressers centuries ago, the ones operating or evoking processes and systems at a distance? The remote controllers. What competencies or positions are required to become a button presser? Who are not allowed to press certain buttons?

Up until the mid-20th century most homes did not have electricity. Before the push buttons became widespread and commonplace, inventors, producers and people in general were grappling with questions about the uses of push buttons. Who should have access to them, what should they be used for and how should they be designed and experienced (Plotnick 2018: 232)? Buttons have been flanked by strings to pull and levers to turn throughout the history of manual control. Which manual skills and postures, which alignments, are fostered by these different devices? What are the choreographies and relations of different manual controls? As part of a mechanical design, the power of a process could be immediately related to the manual power required to initiate the process. Such as a huge lever used to start or to release some powerful physical process. In a world with vast interconnected systems, a slight swipe on a smooth glass-covered screen can initiate way more powerful processes.

The early years of push buttons were also accompanied by hyperbolic promises of ubiquitous automation and an abundance of electrical services. These claims often stood in stark contrast to the somewhat cumbersome everyday experiences of imperfect systems (Plotnick 2018: 230). According to Plotnick, 'a wide chasm existed between the ways people talked about and romanticized buttons for being "simple", "mundane", and "magical" and the terms of their actual use. Buttons often malfunctioned, caused confusion and miscommunication, exacerbated conflict end generated concern' (Plotnick 2018: 230).

When these systems of automation and electricity became more commonplace and reliable, they were slowly mundanized. They became part of everyday life, but increasingly hard to grasp. Since the 1950s, when push buttons started to be implemented at a larger scale, they have become more and more embedded in people's everyday life. Think about all the push buttons that exist in different domestic settings in the 2020s. Which of all these buttons raise concern, are noticed or are the subject of discussion? Once again, who pushes the different buttons? What are the reaches and workings of them? Buttons can be points to start thinking about the ordering of domestic practice. They are also mundane beacons of hidden and often unnoticed systems, mundanized to a high degree. Try a button quest in the place you are in. Buttons are one of the most widespread features of Mundania, small details that come in new guises and variations, as manual command and gesture are employed in new ways.

Get a grip

May 2022. Out in the stairwell where I live, the lights are connected to a timer. On one of the walls of each floor, there is a button. Circular black plastic, with a red glowing light in the middle. When I press the button, the red light turns off, and instead the white light from the ceiling lamps is spread throughout the space. When the button is pressed the timer starts. This system is decades old. Some components have been replaced with new ones. Older components have been combined with newer ones as part of maintenance work. (The word 'maintenance' is also related to hands. The word comes from the Latin *manu tenere*, to hold in the hand.) Without continuous and concrete manual engagement with most systems, components would decay, break apart and fail to function (Jackson 2014; Martínez and Laviolette 2019; Pink et al 2018; Trentmann 2009).

The power button armature in black plastic, with its red glowing light, has a vintage design. It is made to look older than it is. It gives a subtle hint about the age of the underlying system, initially installed in the 1950s. This is made evident when the button is pressed. The system has a mechanical clock that starts to tick. Tick, tick, tick, tick. ... A loud ticking sound is emitted from a meter mounted on the wall of the basement floor. The sound is reverberated in the space. Impossible to ignore. It gives a clear indication that something is going on, a process has been started. Like a taximeter announcing that energy is consumed. Every tick counts. Through sound it announces that the power button has initiated a process, that it has started something. This is, of course, also evident through the function of the system. The staircase is flooded in light once the button is pressed. The light is accompanied by the ticking sound.

I walk out and onto another staircase in the neighbouring house. Here is a similar timer system. The same ticking sounds. I walk down the staircase to my studio and open the door. The sound disappears when I close the door behind me. A music studio is a good place to start examining buttons and manual controls. Among the devices here in the studio is the *Vermona Retroverb Lancet*. It is a small rectangular box, an analog multi-effect processor to be used for sound and music. The small sturdy metallic enclosure has several buttons, knobs and switches. It is made by a small company and workshop in the eastern parts of Germany. It is designed and assembled manually in the workshop. On the company website, Vermona writes: 'Retroverb Lancet offers a comprehensive arsenal of many-sided effects for different applications. The spectrum by far surpasses classic reverberation, filter sweeps or overdrive because you are able to tweak and bend any parameter fast and intuitively to your liking' (Vermona Retroverb Lancet n.d.).

To intuitively tweak and bend parameters, I use the knobs and buttons for manual command and control. The cream-coloured knobs are placed on top of the small metal box. It has an aesthetic that evokes associations with the control panels of earlier industries. Maybe an old powerplant or some of the control rooms of a large factory. There are some aesthetic resemblances to the lighting system on the staircase. Somewhat vintage.

One of the knobs on the box is marked 'cutoff', another one 'resonance'. These are two common parameters to control an analog filter. When the knobs are turned, the sound running through the Retroverb Lancet shifts character. The knobs influence the resistance of components and the currents running through the connected circuits. No digital circuitry or software is involved. It does not house the intricacies of digital equipment, but analog does not mean simple. The combined minute manual movement of knobs can create a plethora of different effects. Sounds can rapidly turn from whisperingly soft and soothing to thunderously loud clamour via such small manual movements: 'even the simplest analogue filters mess with your sound in complicated ways', as it is put in an article on sound design (Reid 1999).

The resonance knob especially has a high sensitivity. Just a slight turn can change the sound dramatically. In some settings the sound turns from a murmur to an ear-shattering squeak. After some practice I have learned how to handle all the knobs. I start to get a grip of the device. It starts to feel familiar. What was first felt as volatile and awkward becomes, in a sense, part of me. The knobs become extensions of my body through which I engage with the inner workings of the circuitry. I can use the controls to turn the Retroverb Lancet into an instrument. A way to manipulate and manufacture sounds.

Buttons, knobs, and switches indicate that below the surface there are circuits and components. The box from Vermona houses several electric components such as resistors and capacitors. The visible controls are also components. Potentiometers and rotary knobs are staples of musical instruments such as synthesizers and effect processors. They have also been common in people's everyday life since electrical devices started to become more widespread. Components have been used to control sound; in other devices they have provided other kinds of manual regulation.

The gradual turning of a knob is a mundane action, but it has some evocative power. In the beginning of the 20th century the radio tuning control knob was introduced. With the radio technology, a slight turn of two fingers holding a knob could tune in radio signals from faraway, evoking electricity and antennas and the interplay between resonance and radio waves. The sounds induced by manual control came into the room through headphones or speakers. Vibrating membranes of speaker elements reproduced the broadcast sound. The manual turning of knobs happened in union and interplay with the sounds from the speakers. Feedback and frequency shifts, resonance and resistance. Rotating knobs and vibrating membranes. Shifts in sounds heard could trigger shifts of fingers on knobs, and vice versa. Listening to radio as well as music production are learned practices that are manual and embodied and involve various technologies (Björnberg 2009; Reyes 2010).

This learning aspect becomes even more evident when the numbers of controls increase. A consumer radio has relatively few possibilities for manipulation, compared with electronic music devices. When starting to use the Vermona device, I had to manually learn how it behaved when I touched it. I had to spend time with it, engage with it to make it mundane, to make me feel comfortable in its presence. I could read the short manual that was shipped with the device to understand some of its workings. But it was when I engaged with it manually that I started to get a grip, when I aligned with it. This is how media and technologies turn into everyday features, how power buttons and rotary knobs become mundane. Through manual and embodied engagement. Some devices require more commitment than others. There is a big difference between the touch of buttons on the power switch controlling the light in my staircase and the Vermona filter. Different conditions to gain results. These devices can be integrated in life in different ways. As instruments to engage with or as simple functional devices that you meet daily and fleetingly.

At some point even the most mundane functional devices have felt new and awkward. Even the power switch in the stairwell has been new and mysterious. Like all fairly advanced technologies and media. In a study of the mediatization of everyday life in 20th-century Sweden, ethnologist Orvar Löfgren underlines how novel and odd the first broadcast technologies were experienced, how they opened up worlds and changed lives:

A farmer remembers his first contact with a radio receiver as an 8-year-old boy, back in 1926: 'I was lent one of the headphones. Oh, Jesus, damn it! You entered directly into heaven as you listened!' … Many have described the magic of the first confrontation with this new media: 'We felt that something new and grand and mysterious had come into our lives … to be able to sit at home and listen to

speech and music from a long distance, it was something of miracle'.
(Löfgren 2021: 40-41)

The listening to a radio could when first encountered be experienced as a miracle. Sound flowed out of speakers. But it was through fingers that control was attained. The turning of the tuning knob on the radio could have an evocative power. It could be both fantastic and mundane. It required some skill but was still quite easy to use. The question is who did the turning and tuning and who listened in different contexts.

The slight shift in position of fingers gradually shifted sounds that emerged from speakers. Static and modulated noise shifted in character and suddenly voices or music could emerge from the crackles and the hiss. Sometimes the sounds faded in slowly or they just occurred abruptly, evoked by a kind of manual surfing of airwaves. Sounds that came from different places, from different transmitters and creators. Solemn voices, cheerful music or dramatic stories. Different soundscapes. The tuning knob not only was a way to capture different radio signals, it also was a way to tune in different moods. Rotary knobs can be not just manual but also affective devices. This manual invocation of different moods, not just information, has gradually become part of everyday routines. Radio introduced new ways to correlate public with private, when 'sounds flowed in and out of the home in ways that did not make the domestic scene more privatized, but turned it, rather, into a laboratory for handling the outside world and organizing social relations' (Löfgren 2021: 38).

The relations between public and private as well as relations between the everyday choreography of manual engagement and the emergence of sensory and affective experiences have shapeshifted when new technologies have been introduced. Radio was a forebear of television, another way to bring new information and sensations, moods and experiences into people's lives. New ways to interrelate bodily moves and manual adjustments to the emergence of new media and technologies. How to turn knobs and how to sit in front of the screen. How to move in a room, how to furnish a room.

At the end of the 20th century the reception of audio visual signals was commonplace in most Swedish homes. Radio and television devices had become staple commodities. The reception of distant electromagnetic signals that were transformed into sensory experiences had become part of everyday life and another set of layers of Mundania was established. Button-based action at a distance was accompanied by sound and video from speakers and screens. The flick of switches and flickering screens, the turning of knobs and shifting sounds that saturated rooms.

Screens have become companions to manual button control in the 2020s. Smartphones, tablets and computers, as well as large-scale screens in public and private spaces, have become so common that it is a hallmark of life in many places where people dwell or roam. It is hard to imagine that before projectors or the television were introduced about a lifetime ago, no screens with electrically generated moving and shifting images existed. Today the screen has become crucial for human experience, communication and sociality in large parts of the world. It is one of the most mundanized technologies around. To manage it we need manual control. Buttons have been around longer than screens.

Fleeting buttons

Screens and buttons permeate Mundania. Often, they have become amalgamated. Pressable buttons appear *on* screens. This was impossible when television first occurred. You could not influence, move, or shift separate elements on the screen of early televisions. The broadcast signal was transformed and experienced, but it could not be altered in any detailed or advanced way. The possibility of manipulating what appeared on screens became publicly available with the first video games. These got mainstream popularity in the 1970s. Arcade games and console games were developed, and at that time devices like joysticks made it possible to shift, move and influence what was experienced as discrete elements on the screens. These possibilities were developed further in coming devices. During the following decades, personal computers (PCs) were equipped with Graphical User Interfaces (GUIs). Now paraphernalia such as the computer mouse were developed. A kind of office variation of the joystick. By using a mouse, a visual cursor or pointer could be moved and positioned over a specific point or area on the screen. When the pointer reached the area, it became ready for engagement. This area could be programmed to function as a button. By pressing a button on the physical mouse, the on-screen button could be pressed, and some process could be initiated. In similar ways, other features on a screen could be manipulated or 'touched'.

These have been the basics of engagement with screens in homes and workplaces for decades, and the idea of screen navigation has become taken for granted. Instructions such as 'move the pointer up to the left, open the menu and press *X*' would have been totally incomprehensible some decades ago. Since the early video games and GUIs, the ways to engage with screens have changed dramatically.

The features of the screen are ephemeral, fleeting, in the sense that a combination of ongoing electric processes must constantly continue if the features are not to disappear. Turn off the power and the features disappear. A fleeting button on a screen is often referred to as being virtual. According to Paul Dourish, the invocation of the virtual has been the central discursive move of digitality (Dourish 2016: 35). This has happened to 'the extent that digital phenomena are rhetorically opposed to non-digital equivalents, and that they further are connected through a notion of displacement, virtual objects – virtual books, virtual worlds, virtual organizations, virtual spaces, virtual meetings, virtual communities, and so on' (Dourish 2016: 36).

On-screen buttons are fleeting and more ephemeral than physical buttons. Often, they are called virtual. But they are not less material. Quite the contrary. Ethereal and fleeting digital entities, rendered as visual elements on screens, are based on intricate material assemblages. The same goes for other systems, based on what is experienced as ethereal or ephemerally ungraspable. Think about radio, wireless networks or devices. It indeed takes a lot of wire to make something wireless. Paul Dourish argues that what is often seen as virtualization, in fact is a rematerialization (Dourish 2016: 36).

As part of Mundania, elements like on-screen buttons are often approached as tangible, ready for action. Despite their virtuality, these might feel extremely concrete. Taken-for-granted parts of the everyday. But the extensive systems maintaining elements like fleeting buttons are often overlooked. This makes a button on a screen something very different from a button mounted on a wall some decades ago, even if they might theoretically have the same function. A press button a hundred years ago could be connected to a widespread distributed system. With contemporary on-screen buttons, the complexity of the system might have increased exponentially. They involve many more devices and much more labour and material than former, simpler systems. This is, however, seldom sensed or made sense of. When a button on a screen is pressed – let's say a button related to AI-, cloud- and web-based services – all the processes that are invoked beyond the screen and the room are hard to sense or to make sense of, and impossible to grasp in their entirety.

Getting the features right

April 2022. I'm sitting by the kitchen table, writing on my laptop. I move the cursor over the background of the Microsoft Word document, using the trackpad of the computer. Slight swipes with the index finger. The smooth surface of the trackpad recognizes the smallest of moves. The cursor appearing on the screen has the shape of an I-beam when it is positioned somewhere in the text area. The beam is an extension of my finger, moving effortlessly according to my intentions. It is a quite pleasurable sensation to see how the beam moves in a similar fashion as my manual movements on the trackpad. It feels smooth.

When I move the cursor to the left or right edges of the document, the cursor shifts guise to a double-oriented arrow. It happens immediately, without any lag. I can move the cursor back and forth: left, right, left, right, beam, arrow, beam, arrow. Here at this specific vertical sliver of the screen, the size and shape of the virtual document can be shifted. Just beyond the edge, the cursor shifts to the guise of another arrow. Now pointing diagonally upward and to the left. Here the cursor has different features and functions. At some parts of the screen, it indicates that now the cursor is located in the Mac Finder. The premises of Apple. I have become accustomed to all these swift graphical changes and shifts of the cursor on the screen. It is part of the taken-for-granted and the mundane. This is how I engage with the tasks I'm concentrated on now. I usually do not think about the shape of the cursor. It is just there, and it shifts appearance in predictable ways.

The very look and feel of buttons and other graphical elements on a screen are extremely important to evoke what could be called 'the right experiences'. Elements, like a cursor, should react as expected. Look, appear, and change according to what is anticipated by a person in a certain situation. Even small discrepancies and deviations from expectations can, however, make something feel awkward, annoying or cumbersome. Here, things might vary. Different people's capabilities and preferences shift. Do you foremost use your left or right hand, how do you experience colours, which size of details and features do you prefer?

Despite variations, some designs have gradually become de facto standards and become widely disseminated, such as the windows, objects and features of computer and smartphone screens. Several of these features have prevailed for decades, while others have changed. New generations of software often come with slight and sometimes more extensive shifts of the screen arrangement. Recurrent adaptation to the changes of software features has become a condition of life in Mundania. Changes are, however, also debated and contested.

A very common design trick is to use metaphors from an earlier and well-known context in a new setting to make it feel more familiar. In software design and on computer screens, concepts such as desktops, canvases, trash cans, folders and windows have (re)appeared to make the new environments feel comprehensible. A specific aspect of this transferring practice is the use of so-called skeuomorphs. A skeuomorph mimics the form, shape and qualities of an object from another medium or the physical world in a new context, like on the screen of a computer or smartphone. According to Wikipedia, 'The term *skeuomorph* is compounded from *skeuos* (σκεῦος), meaning "container or tool", and *morphē* (μορφή), meaning "shape". It has been applied to material objects since 1890' (Wikipedia, skeuomorph, n.d.). The term has now for some time been used to describe interfaces of digital devices and internet-related phenomena (Larsson 2013; Willim 1999).

When it comes to graphical design of smartphones and computers, Apple has been a point of reference for some years. For decades Apple have used skeuomorphic design in their operative systems. The first generations of iPhones and several versions of Mac OS drew heavily on graphical imitations of other materials and textures such as leather bindings of books, paper scraps and green felt. When new devices and systems were introduced, these features were thought to be aids for people to embrace first the personal computer and then the smartphone. Skeuomorphism was used to create a smooth transition for new users, by referring to earlier artefacts, technologies and milieus (Pogue 2013).

Even such a seemingly trivial aspects as the colour of the background on a screen can be an example of design that has some skeuomorphic dimensions. The white background in software like Microsoft Word could create associations with sheets of white paper , white A4s or US letter, like paper on the desks of most offices in previous decades. When the possibilities of inverting screen colours as part of 'dark mode' settings appeared, this associational link was broken. Dark mode was introduced so that users could choose a darker screen appearance to reduce eyestrain when working in poor light conditions, or to simply offer a new aesthetic look. Now it became evident that the background colour of a document on screen had a more complicated and somewhat arbitrary link to the world of white physical paper. This could become obvious when a document including different colours that had been produced in 'dark mode' was printed on a white sheet of paper.

Skeuomorphs have occurred in many settings, and they have had their proponents and critics. With the introduction of iOS 7, the operating system for iPhone, in 2013, the so-called death of skeuomorphism started at Apple. Before this, they had received critique that skeuomorphism had gone too far, that it had become ornamentations without functions (Pogue 2013). The new chief designer at that time, Jonathan Ive, also preferred more minimal designs, avoiding unnecessary ornamentation and references to historical artefacts and products. This preference for minimalistic design could be noticed among several Apple products, such as screens and remote controls. The number of buttons and points of manual control was diminished. As the reduction of skeuomorphs happened, several people were cheering. This is how the earlier skeuomorphism at Apple was described in *The Guardian* in 2013:

> Loosely speaking, skeuomorphism means 'making stuff look as if it is made of something else'. In this context, it is the logic that dictates that Apple's iBooks app resembles a cheap pine bookshelf, for example, and its Notes app resembles a yellow legal pad with lines and a margin – of the type last seen in about 1978.
>
> Look closely, and skeuomorphism is all over Apple and other user interfaces – the little shadows cast by windows, the highlights on virtual buttons designed to make them look shiny, like real buttons. Originally this was to help us neanderthals make sense of the dazzling new technology before us, as in: "Oh, I get it. That looks like a button, so

I'm meant to push it." But Apple got skeuomorphism-drunk, plastering the screens of its futuristically minimal devices with incongruous faux wood, leather and green baize. It got ugly. (Rose 2013)

Skeuomorphs and the shapes and appearance of features, like so many aesthetic choices, aroused feelings. For some, the flat design could be experienced as more tasteful and smoother. Maybe also more future-oriented, with no exaggerated historical references. Questions about aesthetics and good and bad designs and experiences are however seldom simple. Trends and cultural phenomena are never clear cut, in unison and totally homogenous. There was no total death of skeuomorphs in 2013. Skeuomorphism was still important in design in the 2020s. The trashcan, folders and desktop were still design elements in digital environments that clearly referred to physical designs. Several software environments also used other detailed symbols from the world outside computers. But the trend away from skeuomorphism was dominant among several of the major providers of consumer software during the 2010s. In 2019 the neologism *neumorphism* (or neomorphism) also appeared in UX (user experience) circles. Neumorphism was based on the idea of quite a minimal design that took some inspiration from skeuomorphism. Instead of a totally flat design, buttons and other features were designed so that they looked slightly elevated from the background, like embossed structures, giving the impression of depth and 3D without using all the details that dominated skeuomorphism 10 years earlier. With macOS 11, also called Big Sur, Apple was also bringing back some skeuomorphic elements in 2020.

Discussions and controversies about design, aesthetics and how novel designs and technologies should be have obviously occurred before. The discussions about skeuomorphs and about decoration and faux textures and details can evoke associations with early modernism, maybe with the ideas about simplified forms and the rationality and functionality of the German art school Bauhaus during the early 20th century. The omission of unnecessary detail could maybe also be associated with Adolf Loos, an architect based in Vienna and active during the early 20th century. He strongly advocated architecture without clutter and ornamentation. In his lecture from 1910, *Ornament and Crime*, he took thoughts about evolution as the point of departure to explain how humans went through different stages that led them away from primitive ornamentation and towards more sophisticated designs characterized by smooth and clear surfaces (Loos 1908; Canales and Herscher 2005). Loos argued that the progress of culture could be associated with the deletion of ornaments from everyday objects. According to him, it was a crime to force craftspeople or builders to waste their time on ornamentation that would soon become outdated, out of fashion and obsolete.

Design trends and aesthetics do not, however, follow any rulebound evolutionary path, as is sometimes imagined and proposed. There is an ongoing entanglement of reiterations, opposing trends, slow as well as abrupt shifts, and a plethora of parallel preferences and designs emerging, transmuting and disappearing in different social and cultural contexts.

What skeuomorphs can show is that the most minute details can be contested and are often associated with broader ideas about what is and what is not preferable. Ornaments, skeuomorphs or minimal elements are small but important features. The appearance of that which people engage with is a prerequisite for mundanization. When elements appear in what is experienced as 'the right way', they can become taken-for-granted and more or less ignored. This can make life with devices and systems into smooth and comfortable experiences. Buttons are pressed and screens are swiped while people experience that they are doing something else, such as communicating with someone or creating something. The price of the smoothness and comfort is an unawareness of large parts of the things that are used. The shifts and appearances of the most minute details, such as graphical elements of a computer screen, are some of the often overlooked aspects of Mundania.

Interfaces

The graphical appearance of a screen is often referred to as the graphical user interface. Or the GUI, as mentioned in the section, 'Fleeting buttons'. It has become a widespread word, used far away from engineering and design offices. What can be sensed when dealing with a technology is the interface, or the human–machine interface. Buttons, keys, pads, sounds and sensors. Elements shifting on screens, through speakers or through haptic feedback. It glows, moves, buzzes, beeps, fades and swirls. Worn devices, generating layers of mediation between a person and the surrounding world. These seemingly concrete but also evading aspects, interfaces, are at the centre of human–computer interaction (HCI) and the design of UX. Skeuomorphs are part of interface design. Here, ideas about smooth, effective and good experiences of technologies are often at the forefront. How to design suitable and pleasurable experiences? A set of buttons should be organized in ways that appear as logical, users should get comprehensible feedback from their actions, navigation should be smooth, expectations of functions should be met. The user interface is where a human meets or engages with technology.

This is one way to think about interfaces. But there are more open-ended approaches, like the one proposed by Florian Hadler. According to Hadler, an interface is a cultural and a historical phenomenon.

> An interface therefore is not just a surface or a passive gateway or threshold, not only a mode or a site of interaction or communication,

but a deeply historical artifact: a structured set of codes, complex processes and protocols, engineered, developed and designed, a space of power where social, political, economic, aesthetic, philosophical and technological registrations are inscribed. (2018: 2)

Alexander Galloway has furthermore suggested that interfaces are more processes than things (2012: vii). In this sense an interface is not only a meeting point or opening between a human and some manufactured system or assemblage. Not just where the finger meets the surface of the button, and where the pressing of a button generates sensory feedback. Not only some skeuomorphic feature inviting a certain action. Interfaces are entrances to larger issues.

In their analyses of interfaces and their aesthetic dimensions, Christian Ulrik Andersen and Søren Bro Pold have aimed to broaden the concept to comprise not only the designed environment between humans and machines, including buttons, screens and sensors. Neither should the interface be reduced to some of the many contact points and exchanges within a computer or between machines (such as APIs, application programming interfaces). They have proposed the concept of 'metainterfaces' to show how interfaces become ubiquitous and hard to grasp, 'networked or dispersed, at once everywhere, in everything and nowhere in particular as in e.g. cloud computing' (Andersen and Pold 2021: 1). The interfaces become part of environments, ambient and impossible to turn on or off through any simple action. Imagine someone walking into a room with several installed and networked sensors that can register and compute movements, sounds, shifts in temperature and so on. Some technologies might be worn, some might even be implanted under the skin of the person. If all of this is integrated in a dispersed system that provides services, then it is hard to point out where the interfaces start or begin. It is hard to discern what are the boundaries of technologies. When such a dispersed, intimate and ambient system is implemented, it raises further questions about mundanization.

Withdrawal

How are ambient technological systems integrated in public environments as well as in domestic settings? What is expected to be perceived, noticed or become matters of concern (see also Latour 2004)? For whom? The dynamics of attention is crucial for how variations of Mundania emerge. According to Hadler, a well-known proverb among interface designers is that the real problem of the interface is that it is an interface (Hadler 2018: 5). The goal of interface design can be to remove interfaces from attention, to make them disappear, to make them ambient, as Mark Weiser and his colleagues at Xerox PARC put it in the 1990s in their proposition for ubiquitous computing and calm technology (Weiser et al 1999). In a comment titled 'The World is Not a Desktop', Weiser proposed thoughts about the future appearance of computers:

> A good tool is an invisible tool. By invisible, I mean that the tool does not intrude on your consciousness; you focus on the task, not the tool. Eyeglasses are a good tool -- you look at the world, not the eyeglasses. The blind man tapping the cane feels the street, not the cane. (Weizer 1994)

A person who concentrates on the task, not the tool must trust not only the tool per se but all the stakeholders that are involved in the provision of the tool, the technology and the intertwined systems. Anderson and Pold's analyses based on the metainterface also deal with the way that interfaces disappear. The Internet of Things and ideas about smart homes, cities and other places are often centred on the notion that interfaces vanish, and in that sense, they also become harder to grasp. Successful interfaces, however, disappeared long before ideas about smart homes or cities, ubiquitous computing or everyware (Greenfield 2006). When people are concentrating on doing something, they often ignore the devices they use to perform a task. This has also been the case with earlier computer technologies before ideas about calm technologies. When I write this, I do not think about the specific features of keyboard, screen or trackpad of my laptop. This vanishing has to some extent to do with the design of the technologies used, but it is foremost based on habituated action and incorporated practices. Often, people are not attentive to specific features of devices they habitually use, such as buttons or screens. The device (and the interface) becomes like an extension of the body, another entity humans take for granted. The question is when people have to think about the devices they engage with, their bodies, their capacities and the environment they inhabit. People's abilities can shift due to norms and circumstances, and the designs and affordances of things and environments (Egard et al 2022).

When things work as expected, when a device is successfully used for a task or as part of some activity or process, it is not really noticed. Not analysed, scrutinized, or examined. It has *withdrawn*. However, if it stops working, if a process is suddenly interrupted, when the flow is broken, it appears again. Withdrawal has been central to phenomenological theories, and one of the most well-cited examples is the use of a hammer. The experienced user of the hammer does not think about the particular features of the tool, but instead it becomes an extension of arm and hand used to hit nails or other stuff. The world is manipulated, altered and modified by the human+hammer. Anthropologist Tim Ingold, inspired by phenomenological approaches, discusses how skilled practitioners become absorbed in activities. He points out that the carpenter does not inspect the hammer when hammering, not until it misses its mark. Similarly, the musician does not scrutinize the violin while playing, until it goes out of tune, or a string snaps (Ingold 2011: 80-81). Similarly, when I turn knobs on the Vermona device I described earlier, I mostly think about its particular features when it behaves in a way that I have not expected. I may, however, also attentively and reflectively experience it, enjoy my engagement with it, the feel of knobs, its visual appearance, its aesthetic presence. Enjoying it while still concentrating on the task, and not on its specific workings.

The dynamics of withdrawal is in one way obvious, in other ways it is puzzling. It raises deep-ranging questions about presence, awareness, abilities and attention. According to media scholar Shaun Moores, media studies can advance through an increased awareness about these questions, together with a focus on the phenomenological and the non-representational (2021). Moores especially uses the way music pedagogist David Sudnow has written about how skills are acquired in piano playing and typewriting, as well as computer gaming.

Moores suggests that we use phenomenological analysis and a focus on embodied practices and the quotidian to study the role of new digital media technologies (Moores 2021: 65). In his discussion he refers to how Maurice Merleau-Ponty developed a phenomenology about human–thing relations. In *Phenomenology of Perception* Merleau-Ponty also used musicians as an example, and he stressed the importance of embodied knowledge by showing how instrumentalists and musicians such as experienced organ players have embodied knowledge and do not rely on any scheme, analysis or mental map (Merleau-Ponty 1945/2012: 146).

What Merleau-Ponty sketches out is a 'knowledge-in-practice'. Experienced instrumentalists 'have a degree of responsive flexibility, which can be conceptualised as (in a general sense of the term) improvisatory' (Moores 2021: 60). The body of a person has become attuned to things that are used. Bodily postures, gestures and procedures develop through this engagement. Bodies and behaviours are altered through extensive dealings with different devices and technologies, such as cars, smartphones, cutlery, electric guitars, headphones, chairs and sofas. What are the bodies, relations and practices of a tech-infused urban home of the 2020s, or those of a rural peasant's cottage of the 1820s? Or how do technologies merge with bodies in more and more intricate ways, such as digital tracking and monitoring devices and implanted microchips (Fors et al 2019; Petersén 2019)?

Orientations

How is withdrawal related to domestication and mundanization? What can be controlled and comprehended? Tamed? Many technologies withdraw, become part of environments, but are still hard to grasp or comprehend. If a technology withdraws and 'escapes attention', where is it then located? Is the withdrawal a fallacy of perception, or where do technologies go when they withdraw? To play with the spatial metaphor of the whereabouts of withdrawn technologies in Mundania, we could point out possible *orientations*, alignments or emphases of withdrawal (see also Ahmed 2006). These orientations could be used to sketch out ways to further imagine Mundania.

Three proposed orientations could be: in-between, beyond and beneath. These can be used to think about technological withdrawal, the ambient, the distant and the infrastructural. These orientations should, however, not be seen as outright spatial; they are spatio-conceptual, and fuzzy, residing both in space as well as in discourse and imaginaries. We can use them to evoke ideas about the whereabouts of complex technologies, both geographically and metaphorically. Where are crucial parts of systems located, and where do technologies go when they withdraw? The three following chapters will deal with these questions, based on the three tentative orientations, in-between, beyond and beneath.

3

In-between

When media and technologies withdraw, they can seem to vanish into thin air. They become ambient, part of the atmosphere. Ethereal. A kind of technology-induced effortless presence of services and appliances becomes part of everyday life, offering pleasures and conveniences. Oftentimes it is all ignored, just like the elements we live in and with. What are the prerequisites for the technologically influenced ambiences of Mundania, and how do these ambiences merge with the practices of everyday life?

Ambient media

In 1996 Bill Gates, founder and at that time chief executive officer (CEO) of Microsoft, wrote an essay on the Microsoft website. He argued that 'Content is King', which basically meant that the stuff that was produced and conveyed through internet would make a huge difference. Texts and images, websites, even software. This was before social media like Facebook, or Twitter. Before Google. Before the algorithmically engineered feeds of video clips of TikTok, or the flow of images and videos of Instagram. Before the flood of AI-generated stuff, meandering tech-induced conversations, and the bleed of digital inceptions into ever more dimensions of everyday life. In the decades that followed an enormous amount of content emerged through computation and connectivity. Chats, posts, tweets, snaps and ads. Sounds and video. Images and texts. Informal outbursts and calculated campaigns. Informed tuitions and emotive rants. Automated content production. Affective and informational excess.

Over thirty years before Bill Gates' essay about the weight of content, Marshall McLuhan wrote instead that 'the medium is the message' (1964/ 1994). This means that the particular qualities of different media are embedded and in symbiosis with any message (or content). For McLuhan: 'the "content" of a medium is like the juicy piece of meat carried by the burglar to distract the watchdog of the mind' (McLuhan 1964/1994: 18). Since McLuhan, much research has shown that content is hard to separate from media, technologies and social circumstances, but to approach the in-betweenness of media and technologies as part of Mundania, McLuhan's thoughts can be helpful. John Durham Peters stresses some central aspects of McLuhan's media theory:

> Essential are his ideas that each medium has a grammar, an underlying language-like set of protocols for arranging the world and the organs of sensation into a distinct 'ratio,' and that new media can both extend and do violence to ('amputate' was his term) the bodies of those coupled with them. (Peters 2015: 15-16)

McLuhan's definition of media was broad. They are not merely channels of communication or conveyors of information, not just systems, technologies or organizations. They are in-between in a more encompassing way than something that mediates a message between A and B. Media instead create environments. They are transformational. They induce ecologies that become the prerequisite for 'an open-ended transformation in forms of perception and interaction' (Paasonen 2021: 15). In this sense, media have been ambient long before ideas about ubiquitous computing or IoT. When we consider technologies as part of media ecologies and environments, they have the capacity to withdraw into the ambience, to be ethereal, to become in-between.

McLuhan also argued that there are media without any message. His example was the electric light. It normally does not communicate any message, but it facilitates a range of behavioural possibilities (Kitnick 2011). This is media as ambient quality. The light from a bulb can be used to perform several activities. When it is introduced as a feature of people's lives, it gives the possibility to see in darkness. It 'ended the regime of day and night' (McLuhan 1964: 52). It is thereby a facilitator of certain actions, and it is when the light generated by electricity meets other existing patterns of human organization that its energy is released (McLuhan 1964/1994). Then, of course, light can also be used as a medium to convey messages and information, such as morse code, or to indicate something, like traffic lights.

McLuhan's broad arguments about transformative media are characterized by technological determinism, and a high degree of simplification. He argued that electricity and all the media it facilitates would move humanity away from an era dominated by the written word, print technology and dominant ideas about literacy and the role of visual perception. Instead, a new electric age, a kind of technologically induced upgrade of humanity would emerge. What we see today is instead how a plethora of media forms and ways of organization and engagement coexist, in ever more complex and entwined ways. With that said, some of McLuhan's ideas about technological transformation are still useful. Electricity, and the different technologies it enabled, has induced changes of worlds, and it has animated new practices. Humans live in different media- and technology-environments now compared to, let us say, fifty years ago. Life with technologies varies across the planet and between different contexts, but it becomes harder and harder to find geographical and social spots where electric and digital media haven't influenced life in any way. In this sense, media has become something that often resides in-between human relations, and life is encompassed by different kinds of media.

John Durham Peters extends the idea about this in-betweenness of media, about media ecologies and about media that create environments to argue that we should have an even broader definition of the word. He proposes that we see media as elemental. 'Media, I will argue, are vessels and environments, containers of possibility that anchor our existence and make what we are doing possible. The idea that media are message-bearing institutions such as newspapers, radio, television, and the Internet is relatively recent in intellectual history' (Peters 2015: 2). If McLuhan saw roads, numbers, housing, money and cars as media, Peters extends this beyond the humanly created to include air, water, fire and earth. In this sense technologies and media are very much ambient and in-between.

Transient spectacles

May 2022. I sit in the living room. I look at the flickering flame of a lit candle. How different zones of combustion make the flame shift in colour: blue, yellow, and almost white. Its ephemeral but clearly visible edges. Flickering and ever changing. In Swedish it is called a 'living light' (*levande ljus*). I blow out the candle and turn on the electric light. An intricately shaped string of smoke escapes the still-warm wick of the candle. I compare the light from the LED bulb hanging from the fixture overhead to the glow from the candle I just blew out. While the smell of the candle still fills the room, I think about how very prosaic the light bulb feels. So mundane. Functional light. Stable. Something that can be turned on and off. Flick a switch, press a button, and something expected happens. On, off, on, off, swiftly, repeatable. The candle has another kind of aura. It is technologically simple, yet it has certain sensory qualities and an immediacy that can feel compelling. At least in a society permeated by electricity, where candles are not used more than as atmospheric appendages or for ritual purposes.

How do different light sources become affectively charged in different ways, which feelings do they evoke (Strang et al 2018; Bille 2019)? How do the technologies of light become part of Mundania? Electric light can feel mundane and prosaic, but not always. Some years ago, something new happened to electric light in domestic settings. In some places, bulbs started to levitate. This was something new. Strange. Bewildering. In 2016 the Swedish company Flyte was recognized for the best invention by TIME magazine. They had developed a way to use magnetic induction to make things, such as light bulbs or plant pots, float in free air some centimetres over a wooden plate. In the manual for their Flyte light, the company stated that: 'Flyte is a levitating light which hovers by magnetic levitation and is powered through the air. With Flyte, we've set the light bulb free' (Flyte user manual n.d.). The levitating lamp was a way to show off domestic electromagnetism as part of a mundane gadget. To give a prosaic product a new aura. The bulbs came in different designs, some of them evoked the feats of earlier inventors by their names: Edison and Marconi. Some of the bulbs came with 'an Edison style borosilicate glass bulb' (Flyte website).

When McLuhan wrote about light bulbs and electric lighting, he proposed that it is an ambient media that transforms people's sense ratio, influences the way people live, move and spend their time. But ambient light is emitted from different sources, from things and material. When a new technology is introduced, it normally attracts attention, especially if it has some gimmicky qualities. There is a certain tension between what is considered a transformational emerging technology or a gimmick. Flyte's product range makes electromagnetic fields into a spectacle, and the focus of attention is oriented towards the specific features of the levitating thing. These kinds of gadgets are something other than prosaic functional light sources, or ambient light as part of some calm technology that withdraws and 'hides itself'. Instead, the Flyte bulb is supposed to attract attention. To be a conversation piece. Several of the smart home gadgets that were marketed around 2020 used combinations of technologies in unexpected and often gimmicky ways. There were levitating lamps and various ways to use induction and electromagnetism, as well as radio waves, to create spectacular effects. A common trick was also to let glowing digital signs appear in unexpected places, such as inside what seemed like solid wood. Like inside a mui Board, by Japanese company mui labs:

> Designed to improve people's well-being by turning their homes into their oasis, the mui Board serves as the key feature of the mui Platform, our highly versatile platform that enables builders to provide homeowners with unique calm smart-home experiences. … The mui Board turns itself on only when you need to access your important information. The board will otherwise remain off, visually blending into your surroundings as if it's a piece of furniture. The board is designed to let you send and receive digital information through the warmth of the natural wood, changing your relationship with

technology with one look at the panel, or one finger stroke across its surface, at a time. (mui Board n.d.)

These kinds of unexpected functions and appearances were a common trait among developers of IoT devices. The company Flyte also made a product that presented signs inside what appeared as solid wood in their product called Story.

Story consists of a magnetically levitating sphere that orbits around a wooden base telling the time. The base contains a backlit display that illuminates through the wooden surface, giving it a seamless look and finish. For extended features, connect Story to the cloud for using our mobile app and illuminate realtime data through Story's backlit display. (Flyte, Story Walnut n.d.)

There were similarities between the ways mui and Flyte used illuminated signs that appeared in what looked like solid wood. Gadgets such as Flyte or mui Boards were marketed as conversation pieces or as calmly spectacular devices. This 'cool stuff' could turn what looked static and solid into something that was connected to 'the cloud' or could evoke surprise-effects by unexpectedly levitating. Many of these products challenged expectations, but they also provoked questions about appearance and disappearance. They utilized circuitry and electromagnetic fields in different ways, fields that were invisible and unnoticed by humans until they were transformed into various kinds of functions of systems and things. These things were supposed to evoke sensations of wonder, by processes of unexpected automation or appearance.

These are small electric spectacles that can be associated with the ways electricity was utilized to evoke awe and wonder in its early inceptions in society. When incandescent lighting was invented and utilized in the late 19th century it enabled new kinds of illuminations, and it was used for mobile spectacles. In 1881 Thomas Edison organized a procession by his employees, who marched through New York's business district wearing specially designed helmets with a light bulb mounted on top (Nye 2018: 109). The parade of illuminated workers moved down the street while pulling an electric dynamo on a wagon. Each person in the parade was connected to the dynamo through a wire. But to the spectators it looked as if the light-bearing persons moved freely down the street. In the 21st century mobile light, powered by batteries, is nothing spectacular. It is one of the small and mundanized feats of electricity. But light emerging out of wood, or a levitating light, a light bulb shining wirelessly could still fascinate and stir up the imaginaries and expectations of Mundania. At least for a while.

The digital wood board or the induction-induced levitation of Flyte are good examples of gadgets evoking small wonders, gimmicky yet fascinating. They fascinate for a while, until the technologies become either more widespread or outdated because it is not possible to further develop them. For a device or technology to become mundane and to be long-lived it must outlive its gimmick or spectacle phase, it must have some qualities that make it attractive when it has stopped being fascinating.

Flyte's products, like Edison's early spectacles, are built on the interplay between the spectacular and the withdrawn. The effortlessly mobile or even levitation appear to take place in-between, amidst people. But mobile light, as well as later devices that appeared to be ethereal or weightless, requires labour. This labour is often made invisible when gadgets are promoted or shown off. It was not the dynamo on the wagon in Edison's spectacle that caught attention. It was the light bulbs. Neither are the prerequisites that makes Flyte's light bulb levitate centre stage. Nor are the circuitry and components that are mounted somewhere inside a mui Board.

There is always labour connected to devices. Much of it takes place elsewhere, in the production and upholding of systems, something we will come back to. It also takes place as part of everyday use. The time spent on configuring systems and devices is high in Mundania, despite the rhetoric about smartness and ambience of technologies. Some years ago, the devices that are now smoothly used often had to be configured and set up in sequences or careful steps. Now many devices are instead plug and play, or what is called 'class compliant', compatible and ready to use without any specific installation required. But new products that require configuration, consideration and concern keep on arriving. *Configuration time* is a period and state in which technologies do not withdraw, but instead people must engage with them. Where and how does it occur, and how is it related to mundanization?

Gimmicks and emerging technologies at large are seldom implemented as smoothly as it appears in the marketing material. This was, for example, evident during the early implementations of power buttons, as Rachel Plotnick pointed out (2018). For the person buying a levitating lamp, some labour soon became evident. In the box with the lamp, some other paraphernalia were shipped, for an example, a so-called *Flyte co-pilot*, 'a levitation assistant tool' (Flyte co-pilot n.d.) It was a holder made of cork, a set-up device that should be used when turning on the lamp and when trying to get it to levitate. On the website there were tutorials showing how to levitate the lamp. In the manual there was also an 'important note' that stressed that the levitating lamp requires some labour and effort before the magic can happen:

Important Note:
FLYTE was carefully tested before shipment. It is just a matter of practice until you are able to levitate. After about 10 unsuccessful

attempts, it is normal for the electronics inside the base to become warm as you practice levitating.

A temperature sensor will automatically turn off the base if the temperature becomes too hot. If this happens, simply unplug the base and wait 5–10 minutes before plugging the adapter back into the base. (Flyte Instruction Manual n.d.)

The marketing and info material plays with the idea that levitation requires practice and commitment. This is something that is often downplayed when media and technologies are ambient, in-between, calm, or smart. The recurring labour required to get devices to work or to keep them working are seldom at the centre of attention. This labour that a consumer has to carry out in order to get things to work is, of course, merely a small portion of all the effort that it takes by other people to deliver things to Mundania. This labour take place before the parade, the show or the supposed usage of a device. During set-up or configuration time the technology cannot withdraw, it cannot disappear. It is at the centre of attention. Then, however, when devices and gadgets are configured, fixed and doing their work, they can leave the ground, levitate, become ambient, shine and eventually become ignored.

Mundane atmospheres

If technologies and media can be ambient, in-between, then they can also be seen as atmospheric. The words 'ambient', 'ubiquitous', 'immersive' and 'pervasive' all can be related to atmosphere. We could talk about technologically engendered or technology-influenced atmospheres of Mundania.

Atmosphere has been used as a concept within social research and humanities for some time. An influential scholar has been German architectural theorist and philosopher Gernot Böhme. He proposes that: 'Atmosphere is what relates objective factors and constellations of the environment with my bodily feeling in that environment. This means: atmosphere is what is *in between*, what mediates the two sides' (Böhme 2017: 1, italics in original). He acknowledges that the concept has its background within meteorology, but he also says that it has as much to do with practices of scenography. Atmospheres can be produced similarly to how certain moods are evoked on stage, how certain climates or atmospheres are evoked by means of illumination, choice of materials and the appearance on stage (Böhme 2017: 2).

Atmosphere is perceived, sensed and experienced (like the weather), but it can also be produced to evoke certain moods and sensations. There is, however, no straightforward path from production to the sensation (or consumption) of atmospheres. Instead, they come together, they occur, they emerge. According to Shanti Sumartojo and Sarah Pink, atmospheres are induced through activities and happenings in specific spaces or surroundings. Atmospheres are not intrinsic to places or structures. They are emergent aspects of lives that are manifested as people dwell or move through surroundings (Sumartojo and Pink 2020: 6).

Atmospheres do not reside in certain places independent of the persons being there. In that sense they are situational. According to Böhme: 'Atmospheres are quasi-objective, namely they are out there; you can enter an atmosphere and you can be surprisingly caught by an atmosphere. But on the other hand, atmospheres are not beings like things; they are nothing without a subject feeling them' (Böhme 2017: 2). Think again about the person entering a room with networked electronic sensors. What are the different aspects of ambience or atmosphere here? When and how is it activated?

Atmospheres occur when a living creature experiences it in a certain space. They are also affective properties. Atmosphere as a concept pertains to human experiences. The question is how non-human entities may experience atmosphere (beyond the meteorological uses of the word)? How do different animals experience atmosphere? How could a technological system be influenced by atmosphere?

A lot of the promotion and marketing rhetoric around the smart home and technological innovation is saturated with ideas about engineering and design. How to manage and control what is inside a home? The atmospheres of homes? One central aspect of atmospheres is, however, that they are hard to fully control and intentionally manage. Atmospheres can suddenly shift character due to almost unnoticeable variations that are seldom intentionally sparked or instilled by humans.

Atmospheres depend on the material and physical circumstances in specific spaces and at particular times. Anthropologist Mikkel Bille et al state that 'what is needed is a stronger emphasis on the material dimension of atmosphere to balance the anthropocentric perspective on affective experience' (Bille et al 2015: 35). This is what makes the word atmosphere useful in relation to questions about technologies in everyday life. How can the simultaneous ungraspability and very material presence of complex networked technologies be experienced? How do these technologies contribute to the way atmospheres emerge and change? Technologies and devices like voice-controlled speakers or web-connected cameras might influence in several ways that stretch beyond their various technological functions and affordances. To borrow a word employed by Gernot Böhme; these devices are *ecstatic*. Böhme use the Greek word *ekstasis* to describe how "things are radiating into space and thus contributing to the formation of atmosphere. Ecstatics is the way things make a certain impression on us and thus modifying our mood, the way we feel ourselves' (Böhme 2017: 5).

In an article about the orchestration of light among Bedouins in Jordan, Mikkel Bille use the concept of ecstatic things to discuss how green-tinted windows and light sources were used in homes to evoke certain atmospheres, and as part of safeguarding spaces, as part of hospitality practices (Bille 2017: 25). Light is something that radiates, that creates atmospheres and environments. There are some resemblances between McLuhan's thoughts about media that create environments and Böhme's ecstatic things.

Radiation or mood-modification is not something that is necessarily experienced as intense or conspicuous. The ecstasies of things shouldn't be immediately compared to emotional ecstasy among humans. To be in an ecstatic mood or state is an extraordinary experience, to be out of oneself. The atmospheric impetus that comes from things, however, can also be experienced as inconspicuous and mundane. As a hiss or hum in the background, as something that is hard-to-pinpoint or describe. In Mundania this background hum, this atmospheric embrace can be experienced as something ambiguous, as soothing, comforting, but also as creepy, uncanny. The affective charge of atmospheres changes over time; it is dependent on situations and circumstances. This is also how Mundania emerges, comes together, mutates.

Interiors

April 2020. I'm in my kitchen, looking at the computer screen. I raise my head and look out through a rain stained window. There is no one outside. It feels so different to sit here and look out to the street compared to just a few months ago. These are the first weeks of imposed social distancing due to the COVID-19 pandemic. Staying at home had become crucial.

Governments and authorities forced citizens into their homes, during these times of lockdown or distancing. This happened with varying force in different countries. People had to start interpreting what was a rule, a regulation or a recommendation. The virus and the pandemic imposed shifts in practices and also gave a new charge to the word 'home', to the domestic. To the interior. It was outside of homes that risks were to be found. It became a more serious question than before who was allowed inside a residential space. New ways to consider threats in different environments emerged. The mobility and encounters of people and the gathering of publics became obvious risks. Different ways to devise protection and interiority were developed.

One of the central technologies for protection during the pandemic was the respiratory protection mask. The abbreviation FFP (filtering facepiece) had not been part of common knowledge in Sweden before, unless you had been working in environments where you needed dust protection, or for that matter within the healthcare system. In April 2020 you could have a discussion if FFP2 was enough protection against the virus or if you needed FFP3 equipment. Or would it be enough with a medical mask? Aerosol filtration and proper techniques to use this technology became topics of discussion. Is the mask there to protect yourself or to protect others from you? Masks and their different varieties, how to get them, who produced them and how to use them. Face masks as possible fashion items and accessories, white, black, blue or patterned. Face masks as cultural, social and political signs. Who wore them and when? Masks were suddenly part of people's attention, part of debates and popular culture. Part of the atmosphere but also emerging in-between. As shields and barriers. Masks became visible shielding and filtering devices that were raised when people encountered each other or were involved in physical co-presence.

Face masks was one dimension of isolation, appearing between people. Another was the idea about domesticity and The Home. The pandemic home was evoked in popular culture. In April 2020 rapper Drake made a COVID-19-influenced video, based on him being quarantined in his Toronto mansion, while moving to the dance *Toosie Slide*, first popularized through TikTok. In the video Drake is sporting protective gear, like face mask and gloves. The outside world is first shown as empty streets, then mostly present through featured brands like Alyx or Nike, or through references evoked by props, furniture and conspicuous decorations in Drake's house. A quite different musical enterprise, German band Einstürzende Neubauten also made a COVID-19 isolation-inspired video for their April 2020 single 'Ten Grand Goldie', in which vocalist Blixa Bargeld wore a protective medical face mask. The video features footages from: "neubauten.org supporters sheltering in place" as captions in the video shows ('Ten Grand Goldie' (Official video) 2020). The video is full of short snapshots of people's homes, of different interiors, of different atmospheres.

In her research about homes and domestic life, Sarah Pink has stressed how home is more than a space to live in. Home is a feeling. Home is made through people's practices, through relationships, sentiments and what is invested in a place as well as the technologies that mediate how homes are made (Pink 2020). Domesticity can be coupled to gradual processes of settling in, and it is transformed through practices and moods, and how atmospheres emerge and shift.

During the pandemic it became obvious that the way people lived differed. Mansions or tiny apartments, or even the lack of a proper home. The number of persons in a household, and how they got along. Different moods, atmospheres and shifting circumstances. These factors became crucial for how quarantine and social distancing would work. How did people create a place that felt comfortable, homely, safe, and in which isolation was feasible? How were media technologies used in these spaces? Technologies opened up during physical isolation. Chats, video calls, conferencing and online meetings became prevalent.

Technologies can be used both for connection and isolation, which became especially obvious during the pandemic. The attempts to isolate by means of technologies and various kinds of designs during the pandemic have predecessors. Some of them are surreal, such as a peculiar invention from around hundred years ago, 'The Isolator'. This device, a modified helmet, was conceived in 1925 by inventor and science fiction pioneer Hugo Gernsback (Wuthoff 2016; ZKM n.d.).

The Isolator can work as an imaginary companion device to the porcelain insulator that introduced the book. How and when to insulate or isolate, from what, for what? Which technologies and devices become widely used, and which ones remain oddities? The idea behind The Isolator was to create a small headspace, an interiority, in which the user could really concentrate. A sensory enclosure, that was optimized for a person sitting at a desk, reading, and writing. This was an extreme 1925 variation of 2020s pandemic domestic workspaces and rooms redesigned for solitary concentration amongst other inhabitants of homes. It also took the affordances of a mask to its extremes. In Gernsback's invention all outside noise and other sensory impressions were kept at bay by a solid helmet made of wood, cork and felt, resembling the equipment of deep sea-divers. Instead of a diving mask, the user of The Isolator had two circular openings before the eyes, covered by glass.

> It will be noted that the glass windows directly in front of the eyes are black. The construction involved the use of ordinary window glass, the outer glass being painted entirely black. Two small white lines were scratched into the paint, as shown. The idea of this is as follows: The writer thought that shutting out the noises was not sufficient. The eye would still wander around, thereby distracting attention. By having the two white lines scratched on the glass, the field through which the eye can move is comparatively small. In illustration No. 1, it will be seen that it is almost impossible to see anything except a sheet of paper in front of the wearer. There is, therefore, no optical distraction here. (Gernsback 1925: 281)

The small slits through which the user could look were just big enough to see merely a line of text. The helmet was more or less soundproof. This construction should afford extreme focus and concentration. It was an outfit, a device, made for desktop work, but it looked like something for outer space or other exceptionally hostile environments. Outside distractions, such as the hum of traffic or a fly on the wall, couldn't distract, interrupt or influence the person inside The Isolator. Neither could something appearing in the periphery of the visual field distract or intrude on the task performed.

The oxygen to the device was supplied through a hose. The person inside The Isolator got merely what Gernsback thought was needed. A minimal slice of visual information and narrow feedback from the outside world as well as life-supporting oxygen.

Oddities like The Isolator can shed light on what we mean by interiors, and how these are experienced in different times and situations, and by different people. This kind of device raises questions about what the limits of mundanization are. For some persons, an arrangement like The Isolator can feel ridiculous, for others it could be just what is needed. The sought-for device to escape all distractions around. There is no one-size-fits-all design, no atmosphere in which everyone will feel comfortable. What feels ridiculous for some can be perfect for others.

What could be The Isolator of the 2020s, the most extreme measures when it comes to create isolation and interiors? Which kind of interiors are evoked by spatial computing and different wearable devices? What has already become ordinary? Sonic isolation and mediation devices like headphones have become widely popular in public spaces. Headphones and sonic noise cancellation technologies are widespread. When are these technologies considered as appropriate extensions of bodies, and when are they considered as strange and anomalous (Stankievech 2007; Weber 2010; Hagood 2019: 175-180)? Are there other ways to design isolation? From and for what? Extended reality platforms and spatial computing, including new types of headsets are introduced.

How are boundaries raised and managed in everyday life, and what are the borders of atmospheres, what could they be? If atmospheres evoke interiors in which in-betweens can emerge, in-betweens where people and media reside, meet and dwell, what are the outer limits of these? Do different atmospheres have to be separated by recognizable boundaries, or can they have more fuzzy outlines, floating in and out of each other like the conditions of weather? High-pressures meeting low-pressures, cold fronts meeting warm fronts, often stirring up turbulence when they meet. Are the points and slivers where atmospheres meet characterized by turbulence?

Perimeter control

Domesticity can be considered a quite fleeting concept. In many situations, the domestic is, however, often about demarcation. It became obvious during COVID-19 isolation. How is a border, a limit or an imaginary line drawn? Technologies can be used both to challenge demarcations and to instil new borders. The digital extension of walls, when it comes to managing digital flows and connections, is geofencing. The concept implies a virtual perimeter that has an equivalent in the border around a geographical area. When using automation in a smart home-application, for example to turn on or turn off lamps, one routine that could be chosen was to automatically turn on lamps when a resident comes home. For the system to 'know' that the person is arriving home, some device or chip that the person carries must be registered as being inside the perimeter of the home. This device could be a smartphone, that through GPS or when connecting to a wireless network, is registered as being inside the geofence. This is how persons by their very presence can check in to homes and systems. Since the virtual perimeter does not have to follow the walls of buildings or houses, it has a more open-ended relation to geographical and spatial features. The virtual perimeter of a home could extend beyond the physical walls. This again blurs the ideas and imaginaries about what is a perimeter, a home, and what it means to be inside or outside (Steiner and Veel 2017).

When a device carried by a person enters the perimeters of geofencing, it is registered. What are different kinds of registration? To check in to a hotel doesn't mean that you must be physically present within some perimeter all the time you are checked in. You do not have to be in the room to be checked in. You are a temporary lodger, you possess the room and are registered as a guest until you check out, no matter if you are inside or outside the rented room. You have obtained a certain temporary status, which in this case gives you access to some premises. It can be related to the way you log in to a digital service. You are registered in the virtual logbook by the system, and you are logged in until you choose to log out, if there is no set time-limit through which you are automatically logged out after a specific period, maybe due to inactivity. To log in to a digital system means that you register that for some time you have a certain status, that you have access to certain features. With geofencing the check-in/log-in can be synchronized with the geographical whereabouts. This kind of tracking and synchronization was also utilized during the COVID-19 pandemic. Via a specific app, the proximity between different persons could be registered. Who had entered the perimeters around a contagious person? The person inside the area of possible contagion got the status of presumptive carrier of the virus. This person was then forced to isolate. To self-quarantine. To stay at home, to be inside the physical perimeters of domestic walls.

For some years, this play of perimeter practices was part of collective atmospheres. It could be both comforting and uncanny, feel safe or intrusive. Imaginaries about isolation, perimeters, distance and the domestic were combined with digital apps and systems for checking and registering status. Self-quarantine, physical distance, and ideas about perimeters and what should and should not be domesticated were renegotiated and became part of everyday life. Practices and ideas became mundane, but were still often felt as cumbersome and bothering, like an irritating fog in the atmosphere. Certain atmospheric conditions. Banal and strange. A fog that prevented certain actions and that gave an affective tint to everyday life. For a while.

When the pandemic gradually vanished, several of the habits, protocols, concerns and technological operations dissolved, while some remained, such as the increased use of networked technologies for video conferencing, online shopping or e-commerce, as well as various ways to utilize perimeter control. This is how Mundania varies over time while technological dependence also increases.

Attunements

Atmospheres can occur at certain times, in certain contexts, depending on circumstances and on what is and is not present. Technologies, viruses, lighting systems or different people. A way to further explore ideas about atmospheres and about what is in-between, is to use the concept *atmospheric attunements*, as developed by anthropologist Kathleen Stewart. In a writing and thinking experiment she proposes 'an analytic attention to the charged atmospheres of everyday life' (Stewart 2011: 1). She stresses how atmospheres are very concrete but also hard-to-grasp, real and hazy. 'I suggest that atmospheric attunements are palpable and sensory yet imaginary and uncontained, material yet abstract. They have rhythms, valences, moods, sensations, tempos, and lifespans. They can pull the senses into alert or incite distraction or denial.' (Stewart 2011: 1).

Dynamics that Stewart brings up, such as rhythms and lifespans, are central when it comes to an atmospheric approach to mundanization and everyday life with technologies. An approach that appreciates the imaginary and the uncontained, the material and abstract, as well as the automated and the glitchy, the connected and the ungraspable. How are people attuned to atmospheres evoked by such phenomena as automation, isolation, digital registration or perimeter control?

Here, the very local and situated might be linked to more widespread occurrences and to the abstract. Something taking place at a certain time makes some attunements possible, while others are more unlikely to emerge. The material and embodied situatedness is influenced by earlier experiences and memories, by stories heard and by more general currents in time. Just like stories told around a campfire contribute to the atmosphere, widespread imaginaries, conversations and controversies, as well as schemes and campaigns permeate very situated atmospheres. The darkness beyond the enlightened space around the fire is experienced differently depending on the stories heard, just like the experience of the ungraspable in the connected home is experienced in different ways depending on prevailing discourses and fictional stories and worlds. This is one way through which affect and that which is distant can bleed into situated mundane life. This is how atmospheres and embodied presence is enmeshed with imaginaries, with discourses, how the in-between is mixed with the beyond.

4

Beyond

In Mundania, much of what we are connected to is somewhere else. Beyond. *Elsewhereness* has become a normal condition of everyday life with network technologies (Willim 2013b). Occurrences elsewhere have concrete repercussions in our immediate surroundings, and they also influence how we think about such phenomena as places, location and addresses. Systems and devices spread out across the planet influence at a distance and can also be influenced from a distance. Instantly. The assemblages of Mundania stretch out geographically and through imagination.

A manufactured firmament

August 2022. It is beyond midnight and I sit outside. Just outside my home. Looking up to the sky. Waiting for something to happen. This is the time when the comet Swift-Tuttle and the Earth come close to each other. It happens once a year. Like distant acquaintances briefly passing in the night. The comet is followed by a stream of debris. Small particles stretching out in a cloud behind it. The Perseids. That is the name of the cloud. Some of the particles reach the atmosphere of the Earth, creating a shower of meteorites. Dust and stones incinerated as they fall, creating a celestial spectacle. Several shooting stars appear across the night sky every hour during these mid-August nights.

During that night a stronger light appears. At least that was what I heard had happened. I could not see it myself. According to the stories, the light glows brighter than any of the meteorites. Newspapers and television channels start to receive reports from people wondering about the light. Is this really normal? Is it something extra-terrestrial? It looked like a ball, hovering in a cloud, as one person reported to the newspaper *Sydsvenskan* (13 August 2022).

After some hours, an astronomy researcher offered an explanation. It was the light from a SpaceX rocket, launched from California (Karlsson 2022). What looked like a bright cloud, was burnt rocket fuel. SpaceX, the company headed by eccentric and controversial magnate Elon Musk, had launched a Falcon 9 rocket to bring another cluster of satellites of their Starlink programme into orbit. The glowing cloud in the sky, with the ball in the middle, was related to the way the layers of Mundania grow.

Starlink extends electronic communication. It is a network of satellites that should bring low-cost internet to even more parts of the world. To remote and rural areas, as they put it in their promotion material. On the Starlink website a film of the rocket launch could be watched. It was flanked with a rather technical description of the event:

> On Friday, August 12 at 2:40 p.m. PT, Falcon 9 launched 46 Starlink satellites to low-Earth orbit from Space Launch Complex 4 East (SLC-4E) at Vandenberg Space Force Base, California.
>
> This was the 10th flight for this Falcon 9 first stage booster, which previously launched Crew-1, Crew-2, SXM-8, CRS-23, IXPE, Transporter-4, Transporter-5, Globalstar FM15, and now two Starlink missions. (Starlink Mission 2022)

The dry techno-speak about the launch stand in stark contrast to the serene and mind-boggling experience of witnessing the mysterious light in the night sky. Light emitted from burnt fossil fuel, extracted from the depths of the Earth. The remains of biological creatures from millions of years ago. Extracted and transformed, and now burnt by a phallic craft leaving the ground faraway below. Brightly and distantly silent. Launched from the other side of the planet. A beacon from California, creating a spectacle that competed with the astronomical.

The SpaceX event also signalled a growing fact. The firmament is full of devices made by humans. A plethora of satellites for different purposes. Followed by a growing amount of defunct equipment and space junk. The atmospheric layers surrounding the planet are an industrial space. Sometimes these entities in orbit reflect the sunlight and can be seen as artificial stars, parts of a manufactured firmament.

Communication satellites move as small unreachable shimmering dots over the night sky. These faraway entities make local perimeter control and geofencing possible. Holding a GPS device, you can try to imagine the thirty or so satellites that are required to position you geographically. The grid, the system of geographical coordinates, has been sustained by different devices and technologies throughout time. Satellites are prominently remote features that uphold the grid, making advanced logistics and services possible. Mapping, distribution and mobility practices have been transformed. Apparently, also the very firmament has been transformed. The night sky is speckled with human-made small dots. New features have appeared among the constellations of stars. Fodder for new imaginaries. New expanses for the imagination.

The fleeting beyond

What are the perimeters of an imaginary? Are imaginaries individual or social features? What is really an imaginary? Can it be captured, framed or precisely portrayed? Described? There is always a risk when trying to define the formless. Atmosphere can be a useful concept because of its fuzziness. This is also the case with the concept imaginaries. This very fuzziness makes atmospheres and imaginaries rewarding concepts when evoking how Mundania emerges, varies and transforms.

Just like atmosphere, I will therefore let the concept imaginaries stay fuzzy, open-ended and not too distinct and delineated. I will see imaginaries as imperfect entities, as something fleeting that is hard to contain (Willim 2017a). Imaginaries can be both individual and social features. They can be related to extensive constructs like religion, ideology or belief systems. They can be both extraordinarily inconspicuous and the things that makes a difference. They can emerge through imperative practices as well as through acts of daydreaming (Ehn and Löfgren 2010: 123ff). Imaginaries are not always immediately connected to agency, but they might stir up, set in motion, put things in place or relieve, and they can be sparked by certain actions. Imaginaries, like atmospheres, are affective. They might emerge from stories or observations, but they might also emerge through the engagement with beings, devices and technologies.

According to Mark Harris and Nigel Rapport, imagination 'is a common practice, something to which human beings attend whenever they make sense of their environments and situate their life-projects in these environments: a human facility' (2015: xiii). Imaginaries are affective, and they are related to sensemaking. They can be enmeshed in the understanding, organization and comprehension of societies and worlds. The collective sense-making aspects of imaginaries has been promoted through various concepts such as *Social Imaginaries*, and *Imagined Communities* (Anderson 1983). As *Imaginary Worlds* (Appadurai 1997). As *Modern Social Imaginaries* (Taylor 2003). These are all sense-making capacities. Imaginaries that enable practices and the organization of society.

A development of the imaginaries-concept towards studies of technology has been offered by Sheila Jasanoff. She has proposed *sociotechnical imaginaries* as 'collectively held, institutionally stabilised, and publicly performed visions of desirable futures, animated by shared understandings of forms of social life and social order attainable through, and supportive of, advances in science and technology' (Jasanoff 2015: 4). Sociotechnical imaginaries give a kind of direction or orientation in social settings. They are collectively held, to some extent institutionalized, and oriented towards desirable futures. They are normative and point towards what is preferable and what should be avoided. They draw together, unify and give form to the formless. They homogenize.

This homogenizing feature can be criticized as being too streamlined. Too regular. According to anthropologists David Sneath, Martin Holbraad and Morten Axel Pedersen, when an imaginary is defined as something primarily purposeful, as something that can 'fulfil a certain purpose, whether in terms of social function or existential potential' (Sneath et al 2009: 9), it risks losing some of its power. Instead, imaginaries could be seen as something more indeterminate. Imaginaries are not merely plans or strategies. Not merely vison statements, strategic intent or roadmaps. Some aspects of them can be, in Jasanoff's words, 'institutionally stabilised' (2015: 4). But imaginaries also have a more mercurial and escapist daydream character. They might appear to be stable, but they constantly undulate between the directional and the more open-endedly fuzzy. Plans and daydreams blend. Imaginaries are schemes mixed with hope and worries, the strategic wedded to the weird. They are oftentimes ambiguous, mutating, fuzzy and ephemeral.

Aren't imaginaries ambient, or atmospheric? To some extent, yes. They can be locational, but imaginaries reach beyond the atmospheric. They are also characterized by elsewhereness and otherness. They extend the atmospheric past context. They reach into the beyond. They are capacities that are impossible to complete, finish or delineate. They mediate between the mundane and the ungraspable. They bring home wonder and yonder.

The way I use imaginaries when elaborating on Mundania is as this fuzzy capacity, that brings the beyond into the everyday. Here, I am inspired by the way anthropologist Vincent Crapanzano has elaborated on *Imaginative Horizons*. He sees imaginaries as frontiers, as elusive boundaries that never can be transgressed or reached (Crapanzano: 2004). According to Crapanzano, frontiers, unlike borders and boundaries, cannot be crossed. They escape any clear definition. Imaginative horizons make a change in ontological register. 'They postulate a beyond that is, by its very nature, unreachable in fact and in representation' (Crapanzano 2004: 14). He stresses what lies beyond the horizon and the possibilities it suggests, 'the licit and illicit desires it triggers, the plays of power it suggests, the dread it can cause – the uncertainty, the sense of contingency, of chance – the exaltation, the thrill of the unknown it can provoke' (Crapanzano 2004: 14).

Imaginaries can evoke strong, as well as subtle feelings. They can evoke the incredible and marvellous as well as the dreadful or uncanny. At the same time, they permeate the ordinary like a haze, making them related to how atmospheres can be conceived. The difference is that they not only surround, but also stretch towards the indefinite beyond. This elusiveness, the ephemerality of imaginaries, is crucial to the way I evoke Mundania. Imaginaries might stabilize and be part of mundanization, but they also have the capacity to stir up and disrupt, to introduce disturbance in that which for some time has been taken-for-granted. Imaginaries can make the atmospheres of Mundania calm, or they might instil turbulence.

The elusiveness of imaginaries, their powers to nebulously refer to or rather point towards the ungraspable, towards 'something more' makes them related to how narratives might also evoke the strange, the fantastic or the uncanny without really defining it. Hinting at it. Suggesting. As anthropologist Susan Lepselter, inspired by Kathleen Stewart, put it when writing about how people talk about and live with the strange and the uncanny. What is that undefined but alluring 'it' that is referred to in stories?

> *It* is something real, though sometimes the only way to think about it is through its effects in story. The "it," the phantom object of the uncanny or fantastic story, never symbolizes a single "real" thing. Thinking through the uncanny and the fantastic opens up the more general process of how narrative exceeds its literal, referential function to tell a "something more". (Lepselter 2016: 22, italics in original)

The ungraspable become enmeshed with the domestic through narratives and imaginaries. These elusive frontiers have a constant influence on the way Mundania is manifested and how it mutates. Imagination is a certain kind of practice. A social practice (Appadurai 1997). Imaginaries affect atmospheres and are related to other practices and perceptions. Imaginaries might energize or disturb life, and the ordinary, but they will always remain teasingly out of grasp, challenging any attempt at final resolution. They are impossible to outline or exactly portray, they are in this sense unfinished, fleeting, always reaching beyond.

Outreach

April 2022. I have left home. I sit in a car, a small moving cocoon. The sounds of the radio fill the interior. I'm on the motorway E6, by the Swedish Westcoast. I follow the tarmac lines through the landscape, over hills and fields, passing a windpower park. White turbine towers with their spinning blades are erected in rows next to the road. Some kilometres from the towers and the stream of vehicles, the ocean extends westward. The sun will soon set on the horizon. I can't see the shore and the horizon from here inside the car, but I know it is there.

I soon enter a short geographical stretch. From here a line could be drawn westward over the North Sea and the Atlantic Ocean. Straight to the American East Coast. Here, some distance from the coastline, south of the city of Varberg famous for its beach life and its old fortress, more towers are erected in the landscape. Huge antennas rise above the landscape. They also stand in a row and if you follow the orientation of them, you will end up on Long Island and New York City. The antennas have been here since the 1920s. Once they were used for communication between Sweden and the US. From the antennas telegraphic messages could be transmitted to a receiver on Long Island. The signals would have had an extremely long wavelength, up to 17 km.

This is Grimeton. The facility was built around 100 years ago, and the purpose was to have a direct radio link to the US that would help promote business deals and state affairs. It was also meant to facilitate contact between people in Sweden and the Swedes who had emigrated to US (grimeton history n.d.). For some decades it became important for cross-Atlantic communication. Nowadays it mostly has a cultural historical value, even if its transmitter is still functional. Since 2004 it has been a UNESCO world heritage site. This is how it is described:

> The Varberg Radio Station at Grimeton in southern Sweden (built 1922–24) is an exceptionally well-preserved monument to early wireless transatlantic communication. It consists of the transmitter equipment, including the aerial system of six 127-m high steel towers. Although no longer in regular use, the equipment has been maintained in operating condition. The 109.9-ha site comprises buildings housing the original Alexanderson transmitter, including the towers with their antennae, short-wave transmitters with their antennae, and a residential area with staff housing. The architect Carl Åkerblad designed the main buildings in the neoclassical style and the structural engineer Henrik Kreüger was responsible for the antenna towers, the tallest built structures in Sweden at that time. The site is an outstanding example of the development of telecommunications and is the only surviving example of a major transmitting station based on pre-electronic technology. (UNESCO n.d.)

For a while, Grimeton had a central role in sending messages across the Atlantic. In 1994, almost 70 years since the radio facilities outside Varberg were inaugurated, another message was transmitted to reach out over the Atlantic. The Swedish prime minister Carl Bildt sat by his computer and typed a message. He clicked a button and sent an email to US president Bill Clinton. It was the first email transmitted between heads of nations, and the communication got some attention. The message read:

> Dear Bill,
>
> Apart from testing this connection on the global Internet system, I want to congratulate you on your decision to end the trade embargo on Vietnam. I am planning to go to Vietnam in April and will certainly use the occasion to take up the question of the MIA's. From the Swedish side we have tried to be helpful on this issue in the past, and we will continue to use the contacts we might have.

Sweden is—as you know—one of the leading countries in the world in the field of telecommunications, and it is only appropriate that we should be among the first to use the Internet also for political contacts and communications around the globe.

Yours,

CARL (*Internetmuseum, Carl Bildt och Bill Clinton skriver historia med sitt mejlande*, 2014)

Today, when email has been around for thirty years, the Bildt–Clinton email has become a techno-historical curiosity. Part of the history of internet. Also, Grimeton is mostly beyond public attention. Like several other earlier structures made by humans, the antennas just stand there in the landscape, eye-catching, but also to some extent ignored. A witness to earlier crucial occurrences and functions. In the early 2020s, different ways to attract visitors were developed at Grimeton. Guided tours, site-specific artworks, as well as various experiences, such as climbing and an escape room, were presented.

As a functional technological construction, Grimeton is outdated, but email has been highly integrated in daily practices around the planet. It is greatly mundane. The question is how to make that which is established strange again? How can different technologies get the energy to spark imaginaries? How to reach out and spark attention and imaginaries, how to disturb the taken-for-granted and intentionally lift the veil of Mundania? The Grimeton facility is physically present in the landscape, and it can be used to tell several stories and to spark imaginaries. It might also be interesting since it is outdated, and it was never a broadly and publicly mundane technology. If people who did not formerly know about the facilities were to make a detour from the motorway E6 and stand beneath the tall antennas, some fascination and wonder would probably arise. What are these steel structures? The brand new and the obsolete old might spark imaginaries more easily than that which is part of everyday life, since they can be experienced as unusual and strange. Large and imposing buildings, structures and facilities can, due to their very scale, gain attention also after their functions have become outdated. What could be the Grimetons of internet communications of the 2020s? Will there be UNESCO world heritage sites related to email, social media, virtual worlds or generative AI?

Anchor points

Imaginaries often emerge around centres of power and around sites where important decisions and plans are forged. The concentration of power might fascinate. What are then the centres or power, of organizations, of societies? Where is the power most concentrated and how is it made visible? Do modern societies have centres? The question is posed by media scholars

Staffan Ericson, Kristina Riegert and Patrik Åker in the introduction to an anthology about media houses. They draw on media theory when they write about media house architecture as material and aesthetic manifestations of centrality and power, and as edifices that give form to the siteless and the immaterial (2010: 1).

Sites that give form to that which is often elusively distributed can be houses of parliament, main offices (like Googleplex or Apple Park), advanced technological facilities or broadcasting houses (like the TV house in Stockholm, CCTV in Beijing, or the BBC's Broadcasting House in London). The authors discuss how buildings that are hallmarks of media (houses) and organizations throughout the last century can be understood. From TV towers to corporate main offices in Silicon Valley. The tension between the immaterial, the non-localizable and sites of power and control is important.

These constructions can also manifest what Henriette Steiner and Kristin Veel have called gigantism. Which is 'the excessive ways in which the people of modern Western culture have built large structures and thereby brought about enormous unsustainabilities (often masked as progress), vast inequalities (often masked as universalities), and gigantic utopianisms (often masked as meaningful relationalities)' (2020: 4).

How is tangibility of that which is hard to grasp manifested? What are the attempts to breed and direct imaginaries? These questions are important to consider when thinking about mundanization and how more and more complex technologies are incorporated into the layered fabrics of everyday life. I will come back to these questions in Chapter 6, 'Opacities'. What are the anchor points of the imaginations of Mundania, how do they shift over time, and which stakeholders are involved?

Grimeton can help to jam concepts and to shift imaginaries about today's sites of internet and digital cultures. Grimeton was not a flagship building like the TV towers in cities like Berlin or Toronto. No rotating restaurant at the top and not considered to be an anchor point of public imaginaries, attention and the branding of a city, corporation or nation. But it was a facility to reach out that could symbolize the technological endeavours of Sweden at the time. Like many facilities of national interest, such as military or defence facilities, as well as some corporate constructions, architecturally impressive but also paradoxically kept out of public attention (Parks 2009). At least while it was in use. The facility is massive and still evocative, with its metal structures standing above the ground. There is a clear geographical anchor point that can be visited to get a grip on the idea about radio and telegraphic communication a century ago. It is also based on a relatively simple logic, of connections from point A to point B. Not like the rhizomatic and amorphous internet. You can stand below the antennas and imagine radio waves and receiver facilities faraway over the sea. Despite its size, it evokes a sense of comprehensibility. Like a large component in the landscape, graspable, intelligible, roughly like the antenna and tuning knob of an old analog radio. A straightforward functionality, despite its hidden intricacy.

Grimeton can also evoke the beyond in another sense than the geographical. It functions through that which is beyond human perception and sensing. The radio waves, the electromagnetic radiation emitted from the antennas, are not possible to experience without technological equipment. Grimeton is a noticeable construction that might provoke imaginaries about that which is beyond unaided human perception, and it might also shift our imaginaries to the imperative role of antennas in today's interconnected societies.

Electronic devices that can communicate wirelessly, that can send or receive data through air and other media, are equipped with antennas. Smartphones, drones, watches and credit cards. Antennas for GPS, Wi-Fi, bluetooth, NFC (near-field communication) and so on. Antennas are utilized for navigation, shopping, entertainment, phoning and emailing, and they are enmeshed in complex systems that involve a multitude of other components. Antennas and radio waves are crucial components of Mundania that are in-between and that reach beyond.

When antennas and various other components are operationalized, they amount to complex assemblages. Already the technologies and constructions used to send an email in 1994, when Prime Minister Bildt contacted President Clinton, were immensely intricate. The manifold underpinnings of everyday services like email escape attention. It is all distributed, spread out, involving numerous stakeholders and interests. In the 2020s email is considered a well-established technology. Not novel or exciting any more. Not worth press releases or public attention. It has instead become a mundane but essential communication and identification method. Email address together with a password are used and required as identification for several online services.

Given the pervasiveness and importance of email, it is remarkably badly understood, and seldom discussed. It is a lot more incomprehensible than, let's say, the radio communication that took place at sites like Grimeton. If you search for instructions how email works, you will probably get information about multiple servers being involved, about transfer protocols, domains, gateways, clients and so on. It is, however, hard for a layperson to get a good grip on the intricacies of the technologies involved. Neither does email evoke images of any specific edifices or constructions. There are no monumental buildings, antennas or towers of email. No clear anchor points. Instead, email maybe evoke associations with a graphical illustration of a letterbox, a blurry black-and-blue illustration of the internet or simply the sign @. There are no evocative images that stretch the imaginations beyond simplistic illustrations, abstract schemes or technological terminology. Email is experienced as mundane, not monumental, even if it reaches much further than the tall radio antennas of Grimeton.

Handshake

Sometimes, familiar words are used in new ways to grasp the ephemerality of technological processes. When COVID-19 started to spread, people stopped shaking hands. Instead came the Wuhan shake, elbow- and fist-bumps. After some years of physical distancing, people were once again grabbing each other's hands. World leaders could return to the tradition of handshaking, photo-op (photo opportunity) handshakes reappeared and contracts could once again be sealed via the meeting and parting of hands. Can there be handshaking without hands? When people had stopped shaking hands during the COVID-19 pandemic due to risk of contagion, other handshakes increased in volume. The handshakes of machines.

When devices and programs connect in computing and telecommunications, they send signals between them to authenticate, coordinate, negotiate. A process called handshaking. When an email is transmitted, a protocol (like SMTP) uses handshaking to deal with, for example, authentication and encryption.

> A simple handshaking protocol might only involve the receiver sending a message meaning 'I received your last message and I am ready for you to send me another one.' A more complex handshaking protocol might allow the sender to ask the receiver if it is ready to receive or for the receiver to reply with a negative acknowledgement meaning 'I did not receive your last message correctly, please resend it' (e.g., if the data was corrupted en route). (Wikipedia, Handshake [computing] n.d.)

A handshake is not what it used to be. New relations emerge when new technologies and systems are set in motion. New uses of words gradually transform meanings. New imaginaries unfold. When machines handshake it is all about automatic procedures that abolish any manual human engagement. "Handshaking facilitates connecting relatively heterogeneous systems or equipment over a communication channel without the need for human intervention to set parameters" (Wikipedia, Handshake [computing] n.d.).

Device handshaking, virtual handshaking. The worlds of digital technology and computation are full of words that seems to give substance to the ephemeral. Just like the words 'address' or 'domain' orient and direct not only packages but also our thoughts, actions, relations and feelings, the word 'handshaking' turns something impalpable into something more tangible. Or rather, imaginaries about something palpable and concrete emerge.

Concrete clouds

There are also words that evoke imaginaries that turn the tangible into something more ephemeral. The relative short history of networked digital technologies has been filled with these words. Words that can be associated with an aesthetic of ephemerality (Willim 1999). Around year 2000, much that had to do with internet was coupled to the evocative but fuzzy word 'cyberspace' (Willim 2003a; 2017a). It evoked imaginaries of vast digitally engendered worlds. Ungraspable voids. Dark expanses with glowing points of light, data points, servers and routers, like shimmering stars on the firmament. Cyberspace was taken from science fiction, from the short story *Burning Chrome* by author William Gibson (1982). It also appeared in his novel *Neuromancer* (1984). During the 1990's, cyberspace was turned into a promotional hook by businesses and endeavours based on internet technologies. For some years it was used in different contexts and forms. It escaped the worlds of computing and appeared in the marketing of a plethora of enterprises. Cafes, hair salons, schools and pet stores became cyber. Then it mostly disappeared from the discourse, even if the word remained in some contexts. It has continued to be used within security, intelligence and military businesses. Cyber security, war, threats and attacks.

Some years after cyberspace lost its broad evocative power, came the clouds. Networked grids, organized, standardized and programmed structures as well as industrial facilities, were bundled together into more fluffy imaginaries. To transfer a file to server premises located at some undetermined geographical point could be described as 'putting something in the cloud'. These were the imaginary realms where virtual handshakes could take place.

Beyond the lightness of the word 'clouds', there were concrete and significantly heavier constructions and premises. A crucial structural element of the clouds were data centres, even if they were just part of clouds' constitution (Hu 2015). A cloud drive on the desktop of a computer or on the screen of a smartphone pointed to vast connected systems, to one or rather to several interconnected data centres. Often owned by multinational stakeholders, such as Google, Microsoft or Amazon, or by national enterprises like Swedish Bahnhof, these structures became prominent features of the architectures of internet during the first decades of the 21st century. Heavy, industrial, very material, and very energy-consuming, while being promoted as clouds. An ephemeral and airy beyond was in fact a very physical and heavy facility located geographically beyond. Imaginary elusiveness and geographical remoteness were merged.

Artist Matt Parker, who has made an artistic exploration – or what he called a sonospheric investigation – of the sonic dimensions of facilities and machinery from which the cloud was built, has pointed out some specificities of these sites that were not evoked through meteorological words. When he entered the interiors of data centres, he was immediately struck by the rush of air that came up against him (2020: 231). Data centres are closed spaces with a very controlled atmosphere, air pressure and temperature. Inside these buildings, loud white noise was discharged from the machinery. Data centres are far from silent. These clouds are loud and noisy. The facilities produce sonic drones, emitted by cooling fans and ventilation systems that keep down the temperature in the servers and circuits. Listen closely to a nearby server or computer, maybe to some electrical transformer. Listen to the hum of a fan of a computer or some electronic device as it starts to spin to cool down the contained circuitry. Imagine that whirring sound but multiplied thousands of times. How the hum or hiss turns into a roar. Every swipe on the screen of a smartphone, every press on a virtual button of a website contributes to that noise. Smooth swipes at hand and loud noises beyond. Large data centres have large industrial cooling systems, and these are seldom quiet.

When using a smartphone or a computer, when engaging with the clouds, the noise is distant. It takes place on very concrete premises. Sealed, controlled and often undisclosed to the public. The real geography or structure of the clouds involves several stakeholders and locations. Connections, bindings and relations hard to decipher. All the handshakes and contracts sealed that have led to the formation of clouds and to new premises. Where is the data really, how do signals flow, and who or what have been handshaking with what? When these clandestine relations are bundled and branded as "'the cloud' a fluffy and seemingly more coherent imaginary appears. But the clouds of Mundania are the clouds of unknowing, where the user is offered comforting ignorance (Rasmussen 2020). Away from the noisy fans of data centres and any possibly dirty business. Reassuring silence in the presence of convenient and smooth interfaces. Calm technologies at hand. The noise beyond. Distant clamour and disturbing occurrences only appear as quiet hunches through that soothing but also slightly uncanny hum. Sometimes it is noticed among technologies made ordinary.

Clouds, fog, mist

Clouds are very present in everyday life of Mundania. Simultaneously somewhere else, distant, remote, beyond. But sometimes the real edifices of clouds appear nearby. In the early 2020s this could be the result of strategical choices by corporations. Choices about where new facilities were needed due to the whereabouts of customers, or where (green) energy could be inexpensively acquired. It could also be due to geopolitical circumstances and national rules about where data should reside and be managed.

The openings of new data centres and similar large industrial facilities were sought-after investments for municipalities. Places were 'put on the map', people obtained employment, and tax money flowed into local chests. Amazon, Google and Facebook had opened facilities in different parts of Sweden, leading to buzz and debate becoming part of regional identity-building (Vonderau 2018).

When Facebook had established their data centre facilities in the northern Swedish city Luleå, an ice sculpture was erected in the city park. The sculpture was in the shape of a 'thumbs up', resembling the screen icon that was used when someone liked a Facebook-post. Mark Zuckerberg, founder and CEO of Facebook, published an image of the sculpture on Facebook, and in a couple of hours 150,000 people had liked it by using the thumbs–up sign (Sveriges radio 2012). Both Luleå and the data centre facilities gained attention for several days in various places in the world.

The nearby physical presence of clouds is, however, not always something to be desired. It can lead to opposition against disturbing industrial practices being erected too close to people's homes and everyday lives. The NIMBY (Not In My Back Yard) stance often occurs when new structures are about to be built. Windpower mills, battery factories and data centres all can raise negative feelings and concerns for people living nearby the planned facilities. Will the enterprises be noisy or hazardous, will they pollute or disturb? There has always been spatial politics connected to industry and the establishment of large-scale constructions. Bridges, skyscrapers, powerplants, mines, warehouses and factories are contested edifices.

In the middle of the pandemic, in 2021, Microsoft opened a brand new data centre in a small municipality in the southern parts of Sweden. In Staffanstorp, the municipality where Microsoft built their facilities, the expansion of the data centre led to controversies. It was argued that the diesel engines that were supposed to be used if the facility ran out of electrical power would emit levels of exhaust fumes that could be hazardous for the people living nearby. The engines would also be noisy. The business around the establishment of the data centre was scrutinized and questioned through a series of articles in the local newspaper *Sydsvenskan* during Autumn and Winter 2021. In the series of articles called 'The Cloud Factory', the journalists wrote that the process around the establishment and location of the data centre had been clandestine and problematic (*Sydsvenskan*, 11 December 2021).

When a data centre opens nearby, the clouds become present in a new way. The distance of data transmission becomes shorter. Yet what is going on is still, to a large extent, ungraspable. Unreachable and enclosed. Taking place in fenced-off and protected facilities. This has been the case with several constructions and processes throughout the centuries of industrialism. Cloud factories, as well as earlier factories, have been closed spaces. Even in small municipalities where most people could have a connection with a major industry in town, those not working in the factory did not get any insights what took place inside its walls. Factories have been backdrops to everyday life in towns and cities for those not labouring there. Except their visual presence, what could be experienced by those not working in the industries were the outputs. Commodities, goods, smog and noise. Factories and data centres can be close but still faraway (Willim 2005a).

The industrial premises of data centres are the true guises of the fluffy clouds of IT businesses. Data centres, together with all the structures and practices, all the labour and raw material that is required to make data flow, that lets it be stored, processed and compiled. Clouds have also been a computing paradigm, the product of a certain time, a specific type of architecture (Douglas-Jones 2020). In the 2010s, cloud computing was starting to be combined with other architectural concepts and procedures. These concepts were also taken from the realm of the atmospheric, and they extended the arsenal of metaphors evoking ephemerality. Cloud computing was combined with fog computing and mist computing. In a text about IoT development at IBM, Raka Mahesha describe mist and fog computing and give an analogical description.

> The quickest way to understand these two architectures and how they differ from each other is to understand the cloud, fog, and mist phenomena in the everyday, meteorological sense. In real life, a cloud is thick with heavy condensed water hanging high in the sky, far away from the ground. Fog, on the other hand, is the less thick condensed water located below the clouds, and mist is the thin layer of floating water droplets located on the ground.
>
> You can see from that analogy how cloud computing is similar to an actual cloud, where great computing power is located far away from human activities. Fog computing takes place beneath the cloud in a layer whose infrastructure connects end devices with the central server. And, finally, mist computing takes place on the ground, where it is the light computing power located at the very edge of the network, at the level of the sensor and actuator devices. (Mahesha 2018).

Mist, fog and clouds. A new geography of ephemerality emerges. Clouds are farthest away. In data centres, maybe on a different continent. Fog is nearby, maybe as part of domestic spaces, condensed into devices like domestic gateways. Mist resides in the smallest of devices, in cameras, thermostats and sensors. They are all interconnected, they are combined, they all complement each other. The idea is to optimize and to adapt computing to the requirements of new installations, such as large numbers of sensors that register data. This is the world of calm technologies that Mark Weiser had envisioned. Imaginary clouds above and beyond, fog and mist closer nearby, surrounding people as they live their lives. In these combined models, data and computing is in-between and it is beyond. This is to some extent comprehensible, in the sense that these paradigms separate different aspects of computing spatially, give it borders and separate zones. Data is away in data centres, it is at home, and it is dealt with very locally by separate devices and sensors. Three different scales of computing. But how does it all *really* work? The complexity of organizing all these connections, the scales, and all this processing, has increased. It is simplicity, based on embedded and distributed complexity. This increased intricacy must be somewhere. It is to some extent beyond, it is in-between. It is also beneath.

5

Beneath

Large-scale distributed technological systems extend beyond, and they are in the atmosphere, like a fog or like mist. This is all built on multiple layers and arrangements of infrastructure. Technologies of various age, installed for various reasons and with various purposes, are combined. Older constructions and systems influence the way newer systems are built and organized. Infrastructures are hidden beneath streets and floors, inside walls and away from sight. The word '*infra*' in 'infrastructures' is Latin and means below, under or beneath. Spatially, infrastructures are, of course, not always below, but they are often persistently staying under the threshold of attention. Under, infra, sub. Subliminal.

Infrastructures

Sometimes infrastructures are visible. Intentionally showcased. Large factories, constructions and facilities built to function, but also built to impress (Nye 1996; Willim 2005b). Think again about the TV towers in the middle of cities, broadcasting buildings, huge power plants or enormous data centres (Ericson and Riegert 2010). Some of these are prominently visible. Anchor points. Parts of corporations and organizations, as well as parts of infrastructures. Even if these structures may be colossal, they are merely small parts of the systems. Like beacons along wide-ranging routes and networks. Water towers looking like mushrooms, visible fruiting bodies revealing that beneath there are enormously vast mycelium-like, rhizomatic systems of pipes, tubes and ducts. Our societies are full of similar fruiting bodies. Beneath, larger structures are hidden.

Despite these exposed buildings, infrastructures are often not considered. Mostly they are below the threshold of attention for most people. Sometimes they appear as cables and connectors that surface from below (Starosielski 2015). Sometimes they appear as enigmatic constructions, sometimes as points to engage with. Microphones, screens, buttons or maybe small, symmetrically organized holes in walls and sockets, tempting people to insert equipment, to connect and to charge. These are minute yet important points of engagement. Noticeable. Something people might look for. Once again like fruiting bodies, only smaller ones, that capture the attention of mushroom pickers. But the workings, the rhizomatic arrangements past these points are often overlooked and out of reach.

Infrastructures are simultaneously heavily industrial as well as ethereally ungraspable. In this sense they are beneath, but also in-between, often also stretching out beyond. Countless metres of cables and wires facilitate connections and uninterrupted access to the internet and other networks. These connections, these layers, more and more intricate over the years, can go unnoticed because of their ubiquitous and atmospheric prevalence. Cables, devices, and signals transmitted and radiated in the atmosphere. Constructions of various age, sunken into the fundament of society and everyday life. Cables and wirings running through even older subterranean tunnels, whose routes follow the structure of even earlier spatial organizations of cities and other habitats. Newly installed fibre optic cables might run along the spatial structure of medieval streets or beside railway tracks from the 19th century. This is infrastructural path dependency. Infrastructures are, in this sense, sunken into earlier layers of societies.

Infrastructures can have different extension and levels of complexity. They can be 'pervasive enabling resources in network form' (Bowker et al 2010: 98). They can also be small-scale. But when are they noticed? According to Susan Leigh Star and Karen Ruhleder, 'The normally invisible quality of working infrastructure becomes visible when it breaks; the server is down, the bridge washes out, there is a power blackout. Even when there are back-up mechanisms or procedures, their existence further highlights the now-visible infrastructure' (Star and Ruhleder 1996: 113).

Infrastructures are noticed when they break. Infrastructures can furthermore also be experienced in different ways, by different people, when they work. Infrastructures like a barbed-wired wall can prevent some people from moving and acting in certain ways, they can split, divide and make a difference. For other people, the wall can be a shelter, evoking feelings of safety and care. Infrastructures can facilitate or prevent. One of Star and Ruhleder's central claims is that there is no neutral infrastructure. They are concerned with the image that infrastructure is something that impartially sinks into the background of practices. For some, the dividing wall might continuously be a matter of concern. Something that hinders, something that is wished away or even the target for sabotage. For others, it is never noticed, not something to attend to. It stays beneath the threshold of attention in everyday life.

Since infrastructures are not neutral for people, there is no feasible way to distil out the technological from the social. Infrastructures makes a difference. They are not merely neutral backdrops to life, unproblematically distributed along some metaphorical disregarded edges of society. Infrastructures can be experienced and dealt with, felt, and experienced differently by different people, in different situations. Importantly, they influence not only when people intentionally engage with them, but especially when they are not noticed. What is also worth stressing is that infrastructures that prevent, that are oppressive or disadvantageous can become normalized and more or less unnoticed in everyday life. And that which is harmful and unfair can become ignored after a while.

What is needed to sustain infrastructures and make them reliable? Infrastructures must be maintained and cared for to work. Infrastructures seem to be stable, but they are processual, relational and they involve several stakeholders. Earlier public infrastructures have to a large extent become divided between an array of different private corporations and interests (Graham and Marwin 2001). The landscape of Mundania is not only technologically complex, but also organizationally convoluted.

Infrastructures are also relational in the sense that for one person they can be the focus of attention and work, while for another they are something that is ignored, which enables a service or activity, or prevents action in certain ways (Star and Ruhleder 1996: 113, see also Edwards 2019). Maintenance, or infrastructural work, must be recurrent and routinized. In this sense there is an association between material infrastructures and habitual behaviour, between the orderly constructed and regular actions.

Infrastructures are often vast and extensive. They can also be 'mundane to the point of boredom' (Star 1999: 377). They are based on, for example, 'plugs, standards and bureaucratic forms' (Star 1999: 377). Star's argument is, however, far from stating that infrastructures are bland or without meaning; her proposal is instead to carefully study these seemingly boring entities. Because the action is often hidden in that which is overlooked and ignored. What is taking place under the radar can be crucial. The key to new knowledge often hides where nothing seems to happen or where people seem to be 'doing nothing' (Ehn and Löfgren 2010). This is also the standpoint of John Durham Peters, who proposes that: 'Studying how boring things got that way is actually a good way never to be bored' (Peters 2015: 36). He continues to stress that 'the wonder of the basic can beat its banality' (Peters 2015: 37). The basic is the foundational, the underpinning and the support.

Beneath surfaces

August 2022. I grab the smartphone and turn it off. I remove its cover. It feels heavy in my hand. The surface of metal and glass is smooth. I put down the phone on the sofa table, in a standing position. I look at its shiny black glass surface. Its rectangular shape. It is like a replica of the black monolith that appears in Stanley Kubrick's film *2001: A Space Odyssey*. In the film the monolith is mysterious. It has unexplained affordances, but we understand that it has a certain power. It is about evolution, change and about new abilities. Those that engage with the monolith are influenced by it. When they touch it, something seems to happen. There is no direct feedback, but the touch is transformational. Especially in the long run.

I lift the phone from the sofa table and hold it in my hand. I touch the screen and feel the smooth and uniform surface. Every part of the screen feels the same. Then I press the power button, turn it on. After a while, the screen is activated. Now, when I press and touch its different parts, things happen. Visual elements appear and disappear. Processes are started, features are set in motion. Most of the time, my fingertips do not register any difference along the smooth screen. Sometimes, a slight vibrating buzz is felt. But the visual feedback is richer. Objects emerge and withdraw, signs, colours, images and textures come to life.

A smartphone touchscreen is divided into x,y coordinates, that are used to register which part of the screen that is touched. The touchscreen of an iPhone during the early 2020s was capacitive. This means that it uses the conductivity of the human body to induce an electric field between the fingers and the screen. A small electric current flows at the point of the screen that is touched. It sparks action in components beneath the glass (Parisi 2018: 285). Sensors, circuitry and software algorithms transform different touches and gestures into operational data. It is impossible to register the small electrical discharge with the fingertips. Instead, feedback is mostly visual or auditive. Visuals and sounds change and emerge depending on the touches. Sometimes there is also haptic feedback. Small vibrations can be felt through the screen. Buzzes that vary depending on the action of fingers, the position on the screen and which mode the smartphone is in when touched. Invisible actuators mounted beneath the glass of the screen vibrate. Various ways to simulate the feeling of different textures are developed. Various amounts of friction and the imitation of patterns on an otherwise smooth screen are generated by technology installed under the surface.

This illustrates that the points of engagement and everyday dealings with vast infrastructures can be small. 'Though large in structure, infrastructures can be small in interface, appearing as water faucets, gas pumps, electrical outlets, computer terminals, cell phones, or airport security, all of them gates to bigger and submerged systems', as Peters puts it (2015: 30). A small point of the screen of a smartphone can be the actuator of a plethora of widespread processes. And it can vary from time to time. A single point touched on the surface of the small rectangular device can generate a multitude of actions depending on what software is installed and to what structures and entities the phone is connected. And it all takes place beneath and beyond the glass.

It is impossible to notice any difference along the smooth surface if there is no software programmed to generate feedback.

This raises questions about what you really engage with when you use a smartphone. When you put it in your pocket, bag or purse and carry it with you, what do you bring along? It feels like a unified and separate whole. Smooth and unison, like the black monolith from *2001: A Space Odyssey*. A single device. But it is something more. By bringing it along, you are straightaway connected to all that which is submerged, that has withdrawn, that stretches out not only beneath but also beyond. When you take the phone out of your pocket you are aware of its appearance or, more so, what might appear through it. Notices, images, signs, voices or buzzes. Relations, involvements, commitments and bindings. Pleasures. You do not, however, immediately experience the scale of the infrastructure that make notices and buzzes possible, nor all the layers beneath.

Menu diving

In 1995, Microsoft launched a new version of their operative system. Windows 95. It was accompanied by a huge marketing spectacle and event. Stores were opened by midnight on the release date, and the software was sold on Compact Discs (CDs) to customers who had spent a long time waiting in lines. It sold seven million copies in its first five weeks, and it became the world's most popular operating system (Microsoft news, n.d.).

Windows 95 had a startup sound composed by Brian Eno. The hook of the event was the phrase 'Start me up'. Microsoft used the song with that name by the Rolling Stones to promote their brand new offerings. The main thing that was promoted and focused on the event was hidden beneath what came to be called the Start button. When pressed, the button opened the *Start menu*, a graphical menu that appeared at the bottom left corner of the screen:

> Back in 1995, people lined up at midnight to get Microsoft's latest release of Windows, and it was the first version, alongside the enterprise-focused Windows NT 4, to introduce the Start menu. It was designed to make Windows easier to use, and group or organize applications in a list. Before it arrived, Windows users could access apps through Program Manager. It was largely a basic list of apps, with no real organization. While Program Manager did have smaller menus, most Windows users simply launched apps and used it as a list. Windows needed an overhaul. The Start menu was just that overhaul to bring Windows into the next era of computing. (Warren 2016)

A menu was the design that brought Windows into a new era, the next era of computing, as Tom Warren put it, in his 2016 article dedicated to the

Windows start menu. Menus that unfold and give the possibility to reach functions have become a basic design prop in computer software and the visuals of digital cultures. Menus are basic features in Mundania. The practices of *menuing* are taken-for-granted (Willlim 2007). Like button pressing and screen swiping, menuing is a crucial micro-action in Mundania.

Conceptually, a menu of a graphical user interface resembles a physical filing cabinet or a chest of drawers. At least if the drawers are organized in an orderly way. Menus in a graphical user interface contain a set of ordered items. Its contents are hidden until the menu is opened. Inside, there are words, symbols and signs that represent different options. A menu is a feature that organizes choice. Some of these choices can be other menus, positioned inside the main menu. Submenus. Menus are often an indication of a certain level of complexity of a system, where a representational play take place. Items in the menus of smartphones, computers or similar devices refer to offers, which are not present until they are chosen and conjured. In that sense, the menu on a screen is different from the box of drawers. More like the menus used in restaurants. Lists with descriptions of dishes from which to order. À la carte, according to the menu. The practice to look at a list, or in a folder where dishes are presented, to communicate the choice to a waiter who will bring in the ordered dish has been the normal in many restaurants. The name on the list, the name of an item or choice, points towards the infrastructure of the facility. The kitchen where everything is assembled and prepared, where technologies and raw material are engaged. On the computer or smartphone screen the menu option instead points to the hidden workings of software and components beneath the surface of the graphical user interface. Algorithms, software as well as all the (distant) labour and material that is needed to present the offer chosen from the menu.

In the beginning of the 2020's there were several restaurants that offered their menus through apps on smartphones. Here, some different worlds of menus coincide. The infrastructures of restaurants and internet-connected technologies. The menus in the smartphone app and the menu of the restaurant. Look at the menu, position the marker, or the finger on a touch screen, over an object in the menu. Then click to choose it. The machinery will deliver the choice. Things will be set in motion in circuits and network as well as in the kitchen.

Menus offers choices. They also hide and reveal. In software- and user interface design the assortment of functions has often been distributed to dropdown menus, which unfold when they are touched by pointers or fingers. Menus can have several layers or instances, that subsequently roll out or unfold when touched. These menus distribute complexity, an issue that is sometimes within UX-circles referred to as Tesler's law, also called 'The Law of Conservation of Complexity'. It states that for any system there is a certain amount of complexity which cannot be reduced (Laws of UX, Tesler's Law n.d.). Donald A. Norman, an authority when it comes to questions of UX and usability design, uses the example of the automated transmission in an automobile to exemplify Tesler's law. Fewer complications for a driver are accompanied by more complexity in the underlying machinery, 'a complex mixture of mechanical gears, hydraulic fluids, electronic controls, and sensors' (Norman 2010: 46f).

Norman then continues to state that simplicity must always be measured from a point of view. Simplicity for the user means more complexity for engineers. There are also other trade-offs when complexity is distributed to places beneath attention and beyond grasp. The user loses control over the ever more intricate system and must rely not only on engineers, but on all the stakeholders that design, maintain, control, fund, and benefit from increasingly complex and distributed systems.

Scrutinizing practices like menu diving might be one way to think further about the distribution of complexity and the different roles of infrastructures. Menu diving has become widespread and mundane. The choice from menus or ordered assortments is an extremely common practice in societies like Sweden. It occurs in technological interfaces, in consumption and in the ways in which stuff is organized. À la carte and Start me up! Variations of this practice, variations of menuing, are vastly embedded in Mundania. But the questions as to what underpins the systems of menus often remain unanswered.

The hidden library

August 2022. I get the suggestion to access The Library to find the items that I am searching for. This time, I am not searching for books, despite the suggestion to access the library. Instead, I try to find smaller components of them. Fonts or typefaces, such as Garamond, Helvetica or Times New Roman. I am searching for the fonts on my computer, so this takes place on my computer screen. The guide says that the fonts are in a folder in the library. How do I find it?

Devices such as smartphones, tablets and computers have their starting points, where features such as menus are presented. Often a graphical space that opens when the devices are powered up. This is the space where Microsoft's 'start me up' procedure was supposed to take place, where the start menu was introduced in 1995. On Apple Mac computers this space, or instance, has been called the Finder since the Mac was introduced in the 1980's. Here, different graphical elements appear. In the versions available in the early 2020's, there is a so-called dock, where colourful icons for programs and other entities are presented. Then there are folders and, of course, menus. The latter are found at the top of the screen. Hidden beneath names and icons in The Menu Bar.

What's in the menu bar on Mac?

The menu bar runs along the top of the screen on your Mac. Use the menus and icons in the menu bar to choose commands, perform tasks and check status. ... On the left are the Apple menu and app menus. On the right are status menus, Spotlight, Control Centre, Siri and Notification Centre. ... You can set an option to automatically hide the menu bar so it's shown only when you move the pointer to the top of the screen. (Apple Support, What's in The Menu Bar? [macOS Monterey 12] 2022).

In the navigational spaces of computers, some features have been more accessible than others. Like a person entering a building would sometimes have to navigate through a maze-like interior to find their way around, a person engaging with a graphical user interface of a computer in the 2020s had to search through the designs and features of such entities as The Finder on a Mac or The File Explorer on a Windows PC to find what the person was looking for.

Some vital parts of the Mac operative system are to be found in the designated space called The Library. This space, or maybe we should call it a feature, could be found in one of the menus named *Go* in The Finder. When I open the menu, there are several choices, such as *Airdrop*, *Network* and *Tools*. But no library. It seems to be hidden. How to find it? I need a special command to make it visible. The library is a hidden option, or at least a semi-hidden one. To make it appear I must open the Go-menu and then press down the Option-key on the keyboard. Then the word Library appears, next to an icon showing the facade of an imposing house, fronted by pillars. It is a rather banal variation of access. Not that dramatic. But why this semi-hidden option, this secret offer?

Is there something that I should not engage with in The Library? I see the folder named *Fonts* that I was searching for. But here are several folders, such as *Homekit*. It makes me curious. What is in that folder. I click on it and files are revealed. HomeKit is the name of Apple's Smart Home platform. The files have cryptic names. At least they are cryptic for me. There are *datastore. sqlite.shm* and *plain-metadata.config*. The names are incomprehensible for me, and the folder does not feel very inviting, not very homely, despite the name HomeKit. Instead, the names give me a creepy sensation that I maybe should know more about this. What am I supposed to know about this library that I have now accessed through the Go-menu and the option command?

Is this library like the one that Umberto Eco evoked in the novel *The Name of The Rose* (1980/2004)? (Spoiler Alert!) In the medieval monastery in the book, the library is the nexus of a mysterious intrigue. The library is made inaccessible and guarded by the librarian-monk/villain, Jorge of Burgos. The vast and labyrinthine interior space gives place for both knowledge and dread. It houses a poisoned volume, Aristotle's lost treatise on humour. Anyone who touches it and tries to read it will die a horrible and slow painful death. Here perimeter control was not enough. Neither secret code nor sophisticated riddles. The most valuable and dangerous item of the library was instead protected by the librarian Jorge, using the most extreme measures. At the end of the story, the building with all the books is burnt down by the maddened guardian of the library.

The fictitious library in *The Name of The Rose* is something other than Apple's hidden library. But by juxtaposing them, we can evoke thoughts about accessibility, risk, trust and control, beyond the worlds of software architecture and programming. The hidden library (of The Finder) stresses a prosaic logic, namely that there are layers of accessibility and visibility in software design. Ways of distributing or handling complexity, to put it in UX-terms. Once again, it can be compared to the architecture and design of a modern building, where infrastructural features are often hidden. Behind closed, and sometimes locked, doors. Rooms for experts, housing transformers, switches, steering panels and connection points. Equipment often edged by warning signs and instructions as well as cryptic wiring diagrams.

The Library in The Finder is not present upfront because a user should not mess around with vital parts of the system without the required knowledge. The question is, what is required knowledge? Who should be allowed to enter certain rooms? How much does a person have to know about infrastructures and hidden dimensions of systems? Do I have to know what *datastore.sqlite.shm* is to use Apple's HomeKit? Not according to the marketing and guidelines for the platform. But should I know? These are practical questions, but also political ones. Leading to further questions about trust, responsibility and power. Questions that are often forgotten or never raised once complex systems are mundanized.

Infrastructural depth

What is ignored when large parts of systems become infrastructural, and when these constructions grow? The scale of infrastructures is associated with risk. 'The bigger the infrastructure, the more likely it is to drift out of awareness and the bigger the potential catastrophe. There were no train crashes before the railroad was built, and no potato famines before the monocultural overinvestment in that crop in Ireland. Leverage means vulnerability' (Peters 2015: 32). Infrastructures might bring with them unexpected side-effects, or expected and unwanted, but accepted, effects. Like the radioactive waste from nuclear power reactors that must be buried and kept safe for 100,000 years. Waste that is indirectly the product of people lighting up their houses, heating their food or receiving feeds from the internet. Infrastructures are also vulnerable to sabotage, corruption or hi-jacking (Peters 2015: 31). What is needed to hi-jack or sabotage a hammer in a physical workshop, or the hammer in the game on the smartphone? What are the ends of a hammer in these examples? How could different devices be manipulated? How could a home or Apple's HomeKit be trespassed?

As John Durham Peters points out, infrastructures can be large in structure, but small in interfaces, and the vastness is never immediately experienced. However, when assessing risks, the scale and complexity of infrastructures is crucial. To capture the dynamic between infrastructural risk and scale, I would like to use the concept *infrastructural depth*, or infrastructural elevation. People can reach higher by standing on top of multiple layers of structures beneath. More layers give higher reach. New possibilities. With depth and with elevation, however, comes increased risk. To reach high and far, the structures must be large. The larger and more critical the infrastructure, the taller it is, the deeper. The deeper it is, the higher the fall when it breaks or is corrupted. If ways of life are dependent on deep infrastructures, there are also considerable risks.

Large infrastructures, deep infrastructures, are based on the connection and interconnection of systems and things. The question of what should be connected to what must be continually asked. Think about the insulator that featured in the introduction of this book. How should the interplay between connectivity and insulation be handled?

Mundania is built on and with infrastructures. Therefore, the risks of infrastructures are the risks of Mundania. The risks that come with large-scale assemblages, as well as possibly unfair circumstances implanted in structures and processes, are typically ignored when infrastructures become ordinary. Infrastructural depth becomes normalized. This is how mundanization take further hold. This is how the ignorance upon which Mundania is built takes form. This is how the fields of unknowing stretch out, on top of deep infrastructures. This is how also the creepy can become part of the ordinary.

6

Opacity

There can be different orientations of Mundania, such as in-between, beyond and beneath. There can also be different properties. One such property pertains to practices and consequences of concealment. Complex systems, organizations and technologies are hard to oversee and to fully see through. Powerful stakeholders, governments, corporations or other conglomerates can hide their businesses for various reasons. Sometimes justly and fairly, sometimes not. There are also intrinsic trade-offs and distributions between simplicity and complexity, as we saw in the previous chapter, 'Beneath'. Trade-offs between ignorance and knowledge. Between transparency and opacity.

Black boxes

A widely used concept to capture the distribution of knowledge and control when it comes to technologies is *the Black Box*. It has been used as a metaphor or analogy to describe how complexity is hidden in a process, technological artefact, or system. The black box is about concealment of complexity. What does a person need to know to use a device or system, to be part of something, to handle or deal with something? What is possible to know about complex processes and circumstances, and how is this knowledge distributed among different partakers?

The black box, as a concept related to technologies and knowledge, was first used in the development of military technology during World War II (Petrick 2020: 577). The black box as a metaphor was furthermore adopted and transformed within the emerging field of cybernetics in the 1950s and 1960s. Cybernetics grew as a wide scientific field that studied regulatory and complex systems among animals and machines. How could the processes in steam engines, thermostats, electronic circuits, the nervous system or social behaviour be understood through common models involving inputs, outputs and feedback loops? In this sense and context, the black box was a means to analyse systems that were too large or too complex to fully grasp (Petrick 2020: 576).

Since the uses of the black box within military technologies and cybernetics in the middle of the 20th century, it has been brought up in several other contexts. Among these different contexts and associated practices there is also a multitude of understandings of what the metaphor might mean. Maybe this was the case from the very beginning? There is no common understanding of it, which is often true of metaphors and suggestive images that are used in many different contexts. How dark and impenetrable is a black box? In her article about the early uses of the black box metaphor, Petrick notes that it is important to not approach it as a uniform metaphor. To instead destabilize it. To not put the black box in a black box. To acknowledge that it is messy, that it has had many and shifting variants, even in its early uses among cyberneticians (Petrick 2020: 576-577). The question is what analyses and imaginaries the black box metaphor makes possible.

In his widely cited text about the ways in which scientific truths and practices become stabilized and taken-for-granted, Bruno Latour also traced the black box back to early cybernetics. The black box conceals what happens between the input and output of something. 'That is, no matter how controversial their history, how complex their inner workings, how large the commercial or academic networks that hold them in place, only their input and output count' (Latour 1987: 3). Latour stresses that when science become black boxed, what has earlier been evident discrepancies, controversies and even unfair circumstances are concealed. How should the operations that are packed in the box be exposed? One way to unpack what is inside such a metaphorical black box is, according to Latour, to examine how it was once established. To scrutinize what was in flux and at stake before it became taken-for-granted. Different ways to reverse-engineer, or to disclose embedded relations of systems, arrangements and black boxes have characterized much social research about science and technology the last decades.

The question is what the black box metaphor in itself conceals. According to anthropologist Nick Seaver, the black box metaphor 'misrepresents what "access" looks like in practice, and it constrains our methods and the questions we might ask of algorithmic systems' (Seaver 2022: 13). Instead of dealing with questions about the emergence of relationships that define who has got and who has not got access to certain knowledge or processes, the black box metaphor instead encourages us to think in quite a simplified manner about a linear process of input and output where the middle part is simply secluded. 'Conventional ways of thinking about access mislead researchers into thinking that, once they have cracked through the black box's wall, knowledge will be waiting there for the taking. But access is not an event; it is the ongoing navigation of relationships' (Seaver 2022: 15). In reality, the circumstances of secrecy are in themselves convoluted. The algorithmic black box is not merely a technological feat but is instead related to different variations of the social.

Arcane businesses

The black box metaphor has reached beyond only technological systems, and is often used to denote that organizations, means or operations are undisclosed and secretive. When law scholar Frank Pasquale discusses how arcane algorithms and covert surveillance systems are prevalent in a society dominated by a powerful financial sector, he writes about *The Black Box Society* (2015). He paints quite a dire picture of how opacity and concealment is the modus operandi of several stakeholders and systems, especially in the financial world. It is hard to make stakeholders accountable for actions and schemes if their practices are clandestine. He highlights that what he calls the black boxes of reputation, search and finance endanger us all (Pasquale 2015: 18). Advanced technological systems entwined with the opaque operations of finance makes it harder and harder to scrutinize and correct operations and to make persons or organizations accountable (Pasquale 2015: 18).

When large-scale corporations design their financial operations and arrangements, the main objective is not to be transparent and pedagogically straightforward either to the public, to competitors or to tax authorities. What is hidden beyond the upbeat rhetoric of annual and quarterly reports is often intentionally complex and labyrinthine. Undisclosed financial operations are furthermore tightly interwoven with the technologically arcane. Algorithms and technological architectures are often intentionally opaque (Burrell 2016: 4).

What if extensive black-boxing and the arcane are prerequisites for technology-saturated societies and for life in Mundania? Instead of seeing transparency as a major feat of digital cultures, Timon Beyes and Claus Pias propose instead that secrecy and a fundamental opacity is what characterizes much of the tech-based practices in the early 21st century (2019). This opacity is, however, not to be seen as something that emerges from the technological preconditions. Not just a question of interface design and the distribution of technological complexity through cybernetic black-boxing. Instead, they draw on Georg Simmel's essay *The Sociology of Secrecy and Secret Societies* (1906) to propose that secrecy is more fundamental. A central claim by Simmel in the essay is that secrecy is a universal sociological form that is necessary for differentiating social relations (Beyes and Pias 2019: 88). Ideas about total transparency are futile and even dystopic. Total transparency and its discontents recur in several fictional accounts and proposals of dire possible worlds. In works such as George Orwell's *Nineteen Eighty-Four* (1949) or Dave Eggers *The Circle* (2013), the surveillance systems of the evoked worlds suggest absolute transparency when it comes to the activities of citizens and users of technologies while the deeds and intentions of totalitarian regimes or corporate Goliaths remain obscure. The opacity of technology and the social is enmeshed in such a way that they become hard to untangle, all while transparency is hailed. Transparency can be used as a slogan, while it is then often directed towards something and someone. Total transparency is both impossible and when proposed it is often a means of oppression.

In Beyes and Pias's words 'every sociotechnical relation is surrounded by, or shrouded in, a more or less opaque fog of secrecy' (Beyes and Pias 2019: 88). Their intention is not simply to apply Simmel's ideas about the sociological logic of secrecy to new times and circumstances; instead, it is to acknowledge that secrets and secrecy must be taken seriously within cultural analysis. Likewise, I argue that the dynamics of secrecy are fundamental parts in the emergence and transformation of Mundania. Beyes and Pias scramble the ideas about secrecy in modern societies by juxtaposing recent developments with ideas about secrets and secrecy from premodern or early modern contexts. According to them, the treatment of secrets before, let's say 1800 was more differentiated than it is today:

> This is because premodernity was familiar with various types of secrets – such as the *arcana cordis*, the *arcana dei*, the *arcana naturae*, and the *arcana imperii* each of which obeyed different concepts, methods, and rationalities. The primary distinction to keep in mind here is that between the *mysterium* (something nonknowable and thus nonbetrayable) and the *secretum* (something concealed that can be made intelligible and thus be betrayed). (Beyes and Pias 2019: 90-91, italics in original)

The dynamics between that which is non-knowable and what can be made intelligible are shifting over time and context. Which opacities are accepted in different contexts? When it comes to disclosable secrets, the allocation of knowledge and ignorance is important to analyse. Ignorance is, for example, on offer when black-boxing and concealment of infrastructures is implemented (Parks 2009). Selective ignorance can be a unique selling point for a service: 'Don't worry. Just press a button and the rest is taken care of'. Convenience and ease on offer, while complexity is black-boxed. You have, to some extent, to trust (and hope) that complex things made simple are based on fair and safe operations if you want to accept the offer of mundanization. Some, or rather most, people are supposed to not know certain things. We have secret services, warranty void if seal broken, and non-disclosure agreements. Advanced algorithms or financial instruments are not pedagogical endeavours. Corporate roadmaps and strategies are not intended to be communicated broadly. Try to figure out what Apple, Google or Microsoft are to release or do just in the coming month. Secrets are, just as Simmel argued, still a universal sociological form. The question then is what is seen as universally impossible to grasp for any human? What would be the non-knowable mysteries of 21st-century digital cultures? That which is impossible to know. The Mysterium.

Contactless payment

March 2022. I stand by the counter of a local grocery shop. I press a button on the side of my phone. A fast double press. Then I look down on the device. I see a blurry mirror image of my face. The reflection on the screen merges with an image of a credit card. As I look at the screen, sensors register my face. Identify me. A short text appears: "Hold Near Reader". I stretch out my arm and hold the phone close to The Reader. Close to the NFC-reader at the counter. Near-field communication. One of all the protocols for wireless transmission and communication. The phone buzzes. Done! I have paid. A transaction has been achieved. Wirelessly, code was transmitted. Handshaking between machines took place. It happened ambiently in-between, the verification process stretched out beyond, utilized infrastructures beneath what I could consciously comprehend. A plethora of technologies and actors had done their invisible work, to make the phone buzz and to settle the deal. Convenient and concealed, smooth and hard-to-grasp. I took my purchased items and exited the shop. The doors opened automatically before I went out into the chilly air. Contactless payment. A new normal was about to emerge.

The platforms and infrastructures upholding these new payment practices started to appear and become widespread around 20 years into the second Millennium. The services were offered to those who had a smartphone with all the required subscriptions and deals, as well as a valid bank card. These services were based on several technologies, such as ambient wireless communication and facial recognition. The latter was used for this kind of personal identification, and the data of the so-called Face ID on an iPhone was stored in what Apple called a Secure Enclave on the device (Apple Support, about Face ID advanced technology 2022). Facial recognition could, however, also be part of broader applications of operative portraits (Lehmuskallio and Meyer 2022). Databases of face images could be algorithmically processed to instil control and as means for capitalization.

Convoluted simplicity

Facial recognition, various kinds of wireless communication and algorithmic massage came together in contactless payment. New layers of complexity were laid on top of already opaque economic systems. One central process for new acts of payment was tokenization (The World Bank, Practitioner's guide: tokenization n.d.). By registering a card to be used in, for example, *Google Pay, Samsung Pay* or *Apple Pay*, the card became associated with a specific payment system, entwined or layered with the previous payment systems. The card number was replaced with a randomized series of numbers. A surrogate or a proxy for the details on the card (Mulvin 2021). These surrogate numbers were transmitted once a payment was done.

One of the central features of Mundania is how consumption is embedded in ever more complex systems. It is hard to imagine technology separated from consumption in 21st-century Sweden. When technologies become ambient, they are often interrelated with different kinds of economic processes and systems. Consumption becomes ambient. This has been a gradual transformation. Step by step, new procedures and phenomena have occurred. Recurring automatic payments, standing orders, direct debits or withdrawals. A payment method is chosen, then future withdrawals take place automatically, as long as you are in possession of the necessary means and devices. Other transactions are often done with simple gestures that do not involve the transfer of any physical items, nor any tactile encounters. Contactless. This has made economic transactions immensely smooth and easy. Meanwhile, the technologies and arrangements supporting this everyday simplicity has become ever more complex and opaque. This is the paradox of processes such as tokenization as well as cryptography in digital cultures. By encrypting communication or using other advanced systems for concealment of information, new processes become possible. Smooth operations. Ambient consumption. Privacy and security might be maintained through increasingly complex systems. The price is the dependence and trust in opaque assemblages of secrecy and cryptography, with all its associated layers and relations. Tokenization and Secure Enclaves. Complexity is distributed. People must buy into the upholding of the arcane to partake in the systems.

When creating a widespread system for transactions, like Apple, Google or Samsung Pay, a kind of transparency can be offered. The transparency to see what happens with assets and agreements within the service. To get receipts that document all transactions, to see how invoices are settled, how amounts of currency are transferred. What is less transparent is how it all works. If you were to know about the system of encryption or secrecy, it wouldn't be safe. This paradoxical circumstance can be called *cryptotransparency*. Smooth operations and transparent transactions built on intentionally opaque cryptic systems. The encrypted data and technological workings cannot be disclosed if the system is to persist.

Cryptotransparency is a kind of *convoluted simplicity*. Smoothness, convenience, and ease built on complex conditions. This logic is prevalent throughout many dimensions of Mundania. An illustrative example that can give perspective to monetary and financial crypto- and nondisclosure processes is how decluttering software is offered to remove distracting elements from websites and documents. Decluttering on the screen is achieved by introducing more code and components that are hidden beneath and beyond the interface. In the early 2020s there were several plug-ins for browsers like Google Chrome, Firefox or Apple Safari that offered decluttering. And there were apps, such as *Purify*, that remove clutter when a person browses the web. In 2022 Purify was marketed as: 'The simplest, fastest tracking and ad blocker for iOS' (Purify app n.d.). It removed marketing content and streamlined the browsing experience. This is how it was promoted:

> You've got a lot on your plate. The last thing you want is stuff in the way when you're browsing. You don't want a new credit card when you're looking up how old Tom Cruise is. You really don't want an animated picture giving you dating advice. None of that stuff.
>
> So when you're in a shoe store, Purify keeps the taco and lawnmower salespeople out. Because you really don't have time to tell them to go away.
>
> Purify's got your back — and you can finally browse in peace. (Purify app n.d.)

Purify offered a smooth and effective experience. The offer was based on an installation of more software on the used device. Visual decluttering at the price of more complex processes and threads running in the realm of software of a person's device. Tesler's law and the mundanization of more opacity.

There is an organizational equivalent to these cleansing practices and to convoluted simplicity, namely how flaws in an organization are supposed to be diminished by adding different kinds of quality control. Such as organizational programs dedicated to increasing efficiency. An organizational add-on (a team, person, project or division) gets to develop routines to find, measure and collect information about the quality, efficiency and smoothness of ongoing processes. Then employees constantly must report to or relate to this quality control initiative. To assure that efficiency and quality is maintained and hopefully also increased. This is how another layer of complexity is added to the organization. Like the Purify app. Not intentionally cryptic or arcane, yet another add-on that should encourage smoothness or simplicity, while also increasing the complexity.

The plays between opacity and transparency, between complexity, simplicity and smoothness, have been the charms and perils of bureaucracy since it started to emerge. In a text about the relationships between predigital and digital bureaucracy, Barbara Czarniawska elaborates on the intricacies of so-called red tape (2019). She writes about 'virtual red tape' as a new way of framing bureaucratic overflows. The 'red tape' expression comes from the tapes or pieces of string that have been used to bundle (important) documents within organizations, and it has become a metonym for bureaucratic overflow and excess (Czarniawska 2019: 171). Digital technologies and systems are often implemented with the intention to reduce earlier complexities, while the result is more layers of evermore convoluted processes. This is the progress of convoluted simplicity. It is technological, and it is social. Mundania seems to be full of it.

Dreams of transparency

While everyday life with advanced technologies made ordinary is permeated with arcane businesses and convoluted simplicity, the ideas about new technologies sparking and fostering transparency prevail. New technologies for disclosure or clarification are promoted. Often at the price of increased infrastructural depth. Technologies that should unveil and show what is hidden, or that should take away noise and clutter. Like the Purify app, or like various kinds of noise-removal algorithms and other cleansing applications. Or AI-based services that sift immense volumes of data to swiftly offer information. This is how cryptotransparency and convoluted simplicity take stronger hold. Technologies and media as ever more complex devices to unveil reality, that show 'what is really there' have a long genealogy. When Jay David Bolter and Richard Grusin wrote about the logic of remediation in 2000, they proposed that the quest for (visual) immediacy had been a strong current in Western culture (2000). The ideas that new media should offer the possibility to represent reality more transparently can be associated with this current (Bolter 2006).

The price for this quest for transparency is more technological opacity. One response to this has been how several stakeholders have fetishized transparency. Made it into an aesthetic ideal that has not so much to do with exposure, but more with selective showcasing and aestheticization. Some years into the second decade of the 21st century, several large tech-corporations initiated transparency campaigns as part of their public relations efforts. In 2012, Google published suggestive photos that showed otherwise closed parts of their infrastructure. This was part of a transparency campaign that was extended the following years with themes such as renewable energy and efficiency (Willim 2017a). Ten years later some of this material could still be found on a website Google had dedicated to giving selective insight into their data centres. A text stated that: 'Google owns and operates data centers all over the world, helping to keep the internet humming 24/7. Learn how our relentless focus on innovation has made our data centers some of the most high-performing, secure, reliable, and efficient data centers in the world.' (Google Data Centers n.d.)

When navigating the website and the photo gallery, captivating depictions from the facilities of Google met the eye. Several of the photos of buildings were shot during dusk or dawn. Constructions were often coloured by a warm hue, and the artificial lights of the buildings glowed serenely. Several images from inside the facilities showed details of the machinery, pictured in imaginative ways. Many of the images showed long rows of servers and other equipment. Colourful lights, enormous amounts of bundled cables, symmetrically organized pipes, and other structures. Several of the photos did not include any people, while some also pictured people in front of the technological constructions. Employees doing maintenance work or looking with a smile into the camera. Some photos looked like science fiction. They evoked a non-human atmosphere. Machinic. Exactitude, repetition and ingenious designs. Almost otherworldly with all the supposedly advanced but mysterious-looking machinery. Notably many of the depicted constructions were coloured in Google's merry brand colours. Green, blue, yellow and red. It looked both cleverly simple and sublimely complex.

The character of this visualization campaign resembled a visual aesthetic that appeared over a hundred years ago in relation to industries of the 19th and early 20th century. Steel and textile mills were depicted. Long rows of machinery and large constructions stretched out symmetrically towards an imagined horizon. The large scale and the intricacy of machinery was aestheticized, and evoked power and symmetry, as well as well-designed order. Similarly, to the long rows of equipment and personnel in military parades and processions. Symmetry and supremacy. The images were there to evoke fascination for awe-inspiring and carefully designed constructions. Precision coupled with the gigantic (see also Steiner and Veel 2020). David E. Nye has related this visual style and the fascination that was conjured to a phenomenon he calls industrial sublime, or the mathematical sublime (Nye 1996). Industries were showed as imposing and promising facilities, a cornucopia for the future, built on mathematical ingenuity.

When Google published their transparency PR campaign, they inscribed themselves in the history of industrial depiction and promotion. It was also a way to show that ephemeral streams of data and processes of computation were solidly anchored in concrete and imposing infrastructures. Built on advanced mathematics and combined with a tech-utopian stance. Google also inscribed themselves in a specifically American idea, saying that pragmatic mathematics should serve a liberal cause and lead to a better future (Jakobsson and Stiernstedt 2010: 118; see also Turner 2018). The visual aesthetics evoked this idea. The mathematical sublime, or what Nye has also termed 'The American Technological Sublime' (Nye 1996).

The transparency campaign also evoked a certain robustness and anchored ephemeral IT operations in a concrete geography. Places like Dulles in the US, Hamina in Finland or Fredericia in Denmark. This was an intentional way to concretize The Cloud. To point out places where informational flows were massaged. Where algorithms did their computational work and where data was stored. A tangible architecture. The campaign included several websites, not only depicting infrastructures. Technologies and corporate operations were described in several ways, through, for example, podcasts and videos. It was all there to evoke trust and benevolent power, to say and show that what was black-boxed was fair and, in fact, merrily colourful. The campaign was an example of how transparency was used as a buzzword that steered associations away from all the opacities that remained throughout the internet-related industries and in the world of Mundania.

Showcasing and brandscaping

Transparency and visualization campaigns were part of a showcasing practice. But just like a showcase in a museum, they show a selection that has been curated and selected to be aesthetically pleasing. Showcase transparency does not disclose the things and stuff residing beyond the glass boxes, in the warehouses of museums or corporations. Neither does it show infrastructures and operations that take place outside the specific showcase. While Google encouraged people to peep into their data centres through the material on their websites, it was impossible to get access to most of the facilities of large tech-corporations. Instead, some selected places were often promoted as being representative. Headquarters with adjacent visitor centres especially were often shown off. Described in press releases and highlighted in various stories. These areas could include gift shops and photogenic spots that became part of the showcasing of the corporations.

This kind of architectural showcasing has been called *brandscaping* (Klingmann 2007). Brandscapes are places that are chosen to physically manifest the brands of corporations. They appear as flagship stores, such as the Apple Stores that were built in cities around the world. Corporate headquarters can have a similar role. They are often carefully designed aesthetic constructions, built to impress, and geared towards trustwork. Tasteful, and well maintained. Monuments that symbolize trust, that encourage the everyday dependencies on services that uphold tech-infused everyday life. These edifices and the campaigns in which they are shown off reinforce Mundania. Brandscapes are tasteful lures that entice people to start, and to keep on, engaging with products, brands and offerings. When brandscapes are possible to visit, like stores and visitor centres, they are sites where to become part of the brand. Where to identify with the aesthetics. Where to purchase products and where to subscribe to services and become part of corporate worlds.

At many of these sites, transparency was materially evoked. Glass has been a major feat of headquarters and concept stores. Several of the dominating IT companies in Sweden during the early 21st century had their background on the American West Coast. Amazon, Apple, Facebook, Google and Microsoft had all been started and were present either in Silicon Valley in

California or around Seattle in Washington. On the American West Coast were the headquarters. As the corporations became more powerful, they also wanted to showcase their influence through more and more imposing buildings. Through brandscapes. Just like industrialists had earlier created corporate temples and strongholds, new powerful companies manifested their status. With the help of star architects of the time, such as Norman Foster (Apple), Frank Gehry (Facebook) or Heatherwick/Bjarke Ingels Group (Google), brandscapes were created. Apple's circular headquarters, Apple Park in Cupertino, is a case in point. An enormous construction made of steel, concrete and glass. This is how it is described by architects Foster + Partners:

> The simple form of the Ring Building conceals immense expertise and innovation. It comprises a few core elements: communal 'pod' spaces for collaboration, private office spaces for concentrated work, and broad, glazed perimeter walkways – featuring the largest sheets of curved glass ever constructed – that allow uninterrupted connection to the landscape. (Foster + Partners, Apple Park 2018)

Part of the facilities is the Apple Park Visitor Center. Here, visitors are invited into Apple's brandscape. Precision is evoked. A focus on carefully selected materials and elements. 'Nestled within a carefully planted olive grove, an exceptionally transparent envelope sits below a floating carbon-fiber roof, which cantilevers over outdoor seating areas on either side. Its softly-lit timber soffit gives the interior an inviting warmth, while the full-height glazing dematerializes the building volume."(Foster + Partners, Apple Visitor Center 2017)

North from Silicon Valley, in Seattle, Amazon also has used glass as a signature material. In downtown Seattle they constructed glass domes enclosing tropical vegetation. These were called Spheres. Steel and glass constructions on top of concrete foundations, as a kind of office area meets greenhouse:

> The Spheres provide a space to think and work differently, surrounded by nature and the wellness benefits it provides.
>
> The Spheres are a result of innovative thinking about the character of a workplace and an extended conversation about what is typically missing from urban offices – a direct link to nature. The Spheres are home to more than 40,000 plants from the cloud forest regions of over 30 countries. (The Spheres n.d.)

Greenery, wellness and transparency appeared in Amazon's promotion of The Spheres. These were keywords that at the time evoked associations with sustainable and innovative practices. In brandscapes and in the PR campaigns of tech-giants, several affordances of glass were emphasized, such as shininess and translucence. Could this specific material evoke certain powers, could it promote specific processes? Could it be an ecstatic materiality that shaped futures? At least that had been the belief before.

Glass utopias

Glass has been a major component of several spectacular buildings since the 19th century when the material became more widely used. Transparency has often been brought up as an almost utopian feat, and this imaginary seem to recur in slightly different but also familiar guises. An early example of glass architecture, to some extent resembling the ambitions of Amazon's Spheres, was the Crystal Palace. It was built in 1851 for the Great Exhibition in London. In this Victorian building, attractions from different parts of the British Empire were exhibited. With its utopian character, the building had a strong symbolic charge, and it inspired forthcoming architecture and became part of a broader movement through which glass became associated with a bright future. According to media scholar Scott McQuire, the Crystal Palace inspired several of the signature buildings of that time, like the large railway stations that were built in the following decades (2003: 106). Sweeping and curving steel constructions and large glass structures covered the hubs of the new technological systems of mobility. These buildings were 'hailed as the true cathedrals of the nineteenth-century metropolis' (2003: 106). Glass edifices became associated with the future.

During the first decades of the 20th century glass was saluted as a new powerful material. The book *Glass Architecture*, published in Germany 1914 by Paul Sheerbart, influenced architects, writers and critics to propose that glass had a utopian potential. Glass was explained as something that could induce social transformation. The rhetoric about the ways glass could improve society resembles the ways in which new emerging digital technologies are marketed and promoted a hundred years later. Glass and steel structures, or machine learning algorithms. Glass that connects visually and digital network technologies that also connect. Transformative technologies. The ways that critic Adolf Behne argued that glass architecture would wrench Europeans out of their cosy but stifling bourgeoise life are not that different from upbeat visions about revolutionary digital technologies:

> It is not the crazy caprice of a poet that glass architecture will bring a new culture. It is a fact. New social welfare organizations, hospitals, inventions or technical innovations and improvements – these will not bring a new culture – but glass architecture will … Therefore the European is right when he fears that glass architecture might become

uncomfortable. Certainly it will be so. And that is not its least advantage. For first of all the European must be wrenched out of his cosiness. (Behne, cited in McQuire 2003: 107).

Glass was seen as a revolutionary material, offering transparency and social transformation. According to this stance, people must adapt to the futures that are forced upon them by new materials or technologies. Glass should let light into formerly hidden and murky spaces. When it comes to digital technologies of the early 2020s and all the focus on smoothness and convenience, the question is whether the rhetoric nowadays is mostly about people being wrenched *into* and not out of cosiness.

The visions about glass and the power of transparent materials were, of course, part of the broader idea that architecture could transform people and society. It became part of modernist movements, it appeared in the words and practices among influential architects, such as Le Corbusier and the Bauhaus group. The former proposed large often horizontal windows and window walls that would transform domestic spaces and erase the (visual) border between interior and exterior. New architecture for a new society. Walter Gropius (founder of the Bauhaus) wrote: 'glass architecture, which was just a poetic utopia not long ago, now becomes reality without constraint' (McQuire 2003: 107).

These visions about the powerful transformative potentials of new materials, designs and architectures are echoed in the spectacular edifices of corporations like Apple, Google and Amazon. The buildings are like symbolic markers. Brandscapes that connect earlier ideas about architectural transparency and the power of new materials to the present and future product portfolios of the corporations. Emerging generative AI, cloud services, extended reality technologies and new ways to deal with data and information are the new utopian constructions and materials. The glass architectures of the first half of the 21st century are symbolically hinted at in signature buildings of younger corporations. Promotions of corporate futures and new technologies are using utopian visions of the past to gain symbolic authority.

Seeing through

Transparency can lead to insight. That people see through something, that they are not deceived. They see what is inside or on the other side. But what does transparency make invisible? That which someone looks through, that which a person must ignore to be able to see through it. They maybe do not see the glass that encloses something that is showcased. Not until it is smudged or until it starts to reflect that which is outside. Maybe they start seeing their own effigies reflected in the glass, their own faces obscuring or getting merged with that which is showcased.

In this way we can continue to play with ideas about transparency and the affordances of glass, thinking about what is possible to see. Thinking about the variations of transparency and opacity. But there are ends to the transparency concept, limits beyond which the metaphor does not reach. Transparency is a word that comes from the visual world. Something that light shines through. So, what about other sensory modalities? The roaring sonic loudness that artist Matt Parker (see Chapter 4, 'Beyond') was struck by when he visited data centres is not present in transparency campaigns. Thousands of cooling fans and all the air-conditioning equipment that keeps temperatures down are loud. Not to talk about all the back-up energy systems that would start if something unexpected happened in the data centres. During a power outage, roaring diesel engines and generators would start delivering power, also emitting foul-smelling exhaust fumes.

The photos from Google's data centres, as well as the other brandscape depictions from data centres and other buildings, do not give out associations with loudness, foul smells and noise. No emissions of smoke and fumes are evoked or made visible in the material of PR campaigns. No noisy fans or deafening engines. No exhaust fumes. Instead, in the PR rhetoric, renewable energy is brought up. 'Clean', 'green' and 'renewable' are buzzwords that often recur in storytelling and PR among tech-corporations during the first decades of the second Millennium. Serene and almost contemplative stillness is evoked. Or is it not stillness, but rather a kind of smooth high-speed world that is evoked? Not the speed of streamlined roaring modern vehicles. Instead, the silent speed of light of fibre cable connections. Is it a coincidence that the core of these cables is made of glass fibres? The old utopian material once again reappears. Now to offer speedy connections with no noise and no friction.

Smooth silent speed of light. Fibre optic cables and glass constructions of visitor centres and similar brandscapes. The latter built to be aesthetically appealing. Just like many of the devices that are purchased and used. When swiping, clicking, streaming and browsing in Mundania, it is hard to hear any roaring noise. But when following connections, we might come to noisy and dirty places. Many infrastructures and constructions beyond designed product experiences, showcased brandscapes and aestheticized PR campaigns are industrially loud. The noise is elsewhere, but real. The industries that keep up Mundania are built on deep layers that stretch all the way down the drill holes and mineshafts of extraction industries (Parikka 2015; Crawford and Joler 2018). The ghosts of 19th century oil barons are still haunting the planet. Their grip can be increasingly sensed through scorching heat and thunderous storms. The stakes are high, and the infrastructural depth is great. Glass aesthetics, transparency and PR campaigns as they were pursued during the first decades of the 21st century did not encourage people to imagine these circumstances.

7

Order

What role does order play in Mundania? When is this property challenged? Everyday life is where technologies withdraw. Infrastructures and the concealment of complexity are enmeshed in the dynamics of the commonplace, in the interplay between order and disorder. Also uncanny circumstances are veiled below ordered structures and grids, and through often numbing procedures. Sometimes order is challenged.

Standard procedures

Everyday life is rule-bound. It is orderly. Everyday life is however also based on improvisation, the nonsensical and the happenstance. Everyday life is this strange mix of routine and irregularity, where even recurrent disruptions seem to seep into the fabric of acceptance and normality. Uncertainties become enmeshed with monotony. It is paradoxical, 'both ordinary and extraordinary, self-evident and opaque, known and unknown, obvious and enigmatic', as cultural studies scholar Ben Highmore put it (2001: 16). Sometimes things just seem to happen, or they happen while we are busy doing something else. The 'everyday is marked by both repetition and change, and by disciplinary assemblages that not only produce order but also disorder' (Michael 2016: 650).

There is no standard version of the everyday. Yet, standards are enmeshed with the ways life unfolds. A standard is a unifying power in the upholding of technological systems. Technologies and infrastructures are sustained by ongoing processes and arrangements, by agreements and by formalized common understandings. Organization, classification and standards, together with maintenance, induce what seems like a permanent state, makes it possible to experience the variable and even irregular as a solid unified structure. It makes certain things withdraw while others pop up and provoke attention.

Standards, classifications and categorizations can be visible, and they can be manifested physically, like in the row of numbered architectural box-like houses along a street. Processes of classification are also part of the infrastructural, through withdrawn systematicity. The numbers on the houses, the addresses are indications of a drawn-out expanse of bureaucratical organization, classification and ongoing procedures. Population registration and utility services, statistics and connectivity.

These aspects of ordering are often invisible or discreet, while simultaneously being regulatory and normative. Classification appears in everything from the spatial organization of a city or a home, to the appearance of elements on the screen of a smartphone or the procedures of bureaucracy or the routines of a household. Part of lives, yet 'ordinarily invisible' (Bowker and Star 1999: 2). The outcomes of classification, or the even more disciplined system of standardization, are experienced, but the organization often resides in the background.

Standards might sometimes be felt, especially when they are challenged. When you bring an adapter that does not work with a device. When you have the wrong batteries or cables. When you have two devices that do not support each other, that can't perform the handshake. When you wait for the delivery of a parcel at the wrong address or at the wrong time. What if you do not have an address? Not on the population register. No passport. No smartphone. Then you are outside the standardized meshwork that keeps most societies together. Offline, off the grid, off the record.

Now and then, something happens that upset orders or that introduces new ones. Old standards might become obsolete, and new standards might be introduced. One such shift became evident during the COVID-19 pandemic. To approach and try to pass a border control without a negative virus test result that had been produced according to specific routines and procedures, or without the right certificate, signatures or QR codes, could be a distressing experience. New concepts become part of vocabularies and routines. This kind of standardized and ordered system becomes clearly noticeable and felt when it is challenged. Then, as the pandemic receded, many of the orders vanished again. As a citizen or user of systems, you must keep yourself updated with the shifts of standards. Know what is expected and compatible. Carry the right certificates and documents, or the right devices and adapters.

Suppressed accessories

August 2022. A large heap of white plastic. Some parts are shiny, some stained. A salad of wires and a plethora of adapters. Some are broken, showing their interiors. Thin threads of metal and soldering. I've decided to dig out all the white-coloured Apple accessories that have accumulated during the last decades. They are associated with different standards and protocols. Some outdated, such as the 30-pin connector, DVI, RCA, component video, Magsafe 1 and 2, or Thunderbolt 1 and 2.

This heap of mostly derelict stuff is one side of the wireless and the smoothly designed streamlined aluminium and glass universe of one of the world's most powerful providers of consumer electronics and services. The corporate road of Apple is lined with a trace of derelict, mostly white-coloured cables, converters and accessories. The same goes for several of the producers of electronic consumer goods, even if most companies haven't been so consistent in keeping a unison colour-code and design style among this often-overlooked range of products.

New digital technologies are often promoted as weightless and smoothly organized. Like the electrical machines of early 20th century were promoted as something that would remove messiness once and for all by introducing new machinic efficiency, new technologies are never promoted as muddled and messy, or as something introducing new kinds of disorder. But when looking at all the things that accompany flagship devices, we find a, seldom well-ordered, world of connectors and peripherals. Accessories and cables. These are often overlooked when discussing emerging technologies. Yet, this paraphernalia recurrently pops up. Often irritatingly present in its relative unmanageability. They are like snaky, yet vital, umbilical cords to all the vast hidden infrastructures that stretch out beyond and beneath. Often too palpably twisty to be associated with either cloud or grid.

Cables, wires and connection points are often intentionally concealed. For every cable and cord, a cable canal seems to be designed. Ways to hide the mess are promoted by home improvers, handypersons and designers. Like the US TV personality Bob Vila, who on his website presents some tricks to hide wires and cables. Hook them to the back of your furniture, corral them behind the couch. Buy a cord cover, stick it to the wall, put the cable inside and paint the cover in the colour of the wall. Slip wires into a drawer or snake them through baseboard accessories (Reddigari 2022). In a consumer society there is often at least one more accessory provided for every accessory. A product or service to solve a problem instilled by another product. This consumption, configuration and design game is intrinsic to the way Mundania unfolds in the 2020s. Processes of consumption geared towards concealment are part of almost ritualistic suppression practices when it comes to electronic devices.

Ordinary mess

Outdated accessories are often associated with abandoned standards and formats. Standards to connect devices, like USB, have been relevant for some time. Others, like Firewire, that were widespread at the turn of the century, are obsolete in many contexts. ISO standards are revised at least every five years, and other standards change as technologies and the world change. Standardization can be related to how products and procedures are incorporated in various kinds of frameworks for sorting, identification and verification (cf. HaDEA 2023).

Legal verification frameworks and standards, as well as classifications, incite order and make differences. They orient work and attention, and steer what is possible. For better or worse, classification is never neutral. Never in stasis. Classification is something that goes on, that emerges and recedes. In that sense the word 'standard' is a bit trickier. It evokes the idea that it should be static and stably constant. But a standard is not unchanging. Its mutation just occurs at a relatively slow pace. Not even the most widespread systems, like internet protocols or the metric system are forever, and many are apparently not globally applied.

Even when things are compatible and verified, even when devices and systems are in sync with standards and classifications, they can be experienced as messy. Cable- and wire-salads, and ways to hide them, as well as all the practices that come along with this, are good entry points to think about the ways experiences and discourses of standardization and messiness change among situations, persons and different social settings. When is something experienced as mess and subsequently hidden? How do variations between pedantry and sloppiness unfold, and what is the role of different technologies? What does minimalistic designs hide? What do ambient technologies hide?

Since mess is often seen as the evil twin of order, people seldom brag about messy lives or homes. Instead, mess is pictured and said to be a temporary anomaly. Some people welcome guests into their homes apologizing for the supposed momentary mess. The truth is often, however, that flawless order is a very transitory state (Löfgren 2017: 1). Then again, for some, in some situations, it is perfectly OK to be a bit messy. For whom and when is this OK? It can even be associated with creativity and an open-minded bohemian lifestyle. The question is when and how positive and progressive messiness turns into something that is experienced as negative and unfavourable. Which roles does mess have in Mundania? When and how is tech-related complexity experienced as mess? Order mostly works as part of a play of contrasts and comparisons. 'Why Do Things Get in a Muddle' (Bateson 1972/1987: 13-18). 'My order may be your mess. Differences of class, gender, ethnicity, and generation are at work here. The production of disorder is a cultural practice' (Löfgren 2017: 1-2).

In some situations, even slight mess can make people nervous. Could the irregularities be the spark that ignites an explosive negative process? From clutter to chaos. When everyday life and its material manifestations are experienced as unbearable messy, certain practices and rituals can be mobilized to instil order. Cleaning, tidying up, and the management and maintenance of classifications and systems. The emptying of mailboxes and the organization of files. This is when cable salads are concealed or disentangled, or when the array of different accessories is organized. Power adaptors in one box, network cables in another and so on. This is also when some things are stowed away to eventually be discarded. After a while the discarded might gradually turn into mountains of waste. Elsewhere. Dark matters that mount up beyond the orders of Mundania.

Technological order in Mundania is furthermore often associated with different corporations and the way they launch products and systems. The interplay between disorder and order is often made visible when a new system is introduced, or when different systems clash. At the fringes of corporate systems incompatibility often occurs. A result of different stakeholders trying to dominate markets or influence the ways technologies should be developed. In the early 2020s a cable for an iPhone was not compatible with a phone from Samsung, and vice versa. This incompatibility is just the most obvious example. Incompatibility and mismatches between systems can be clearly noticeable and even dramatic, while at other times they are experienced as more modest frictions, glitches and lags.

Snap to grid

August 2022. I am about to open a document in Microsoft Excel, the most well-known spreadsheet software. At least at the time of writing. Here be choices. Menuing. Make a household monthly budget, a loan amortization schedule, a credit card tracker or a grocery list. These are some of the available templates that are on offer when I look at the screen. The software has several lookalikes, such as Apple Numbers and Google Sheets. The templates show prefabricated designs of the different functions. Figures and words can be changed according to preferences, purposes and needs. It all fits into the grid that underlies the logic of spreadsheet software. The more advanced the software has become, the more colourful and intricate the looks of the graphic appearances of spreadsheets. But beneath, there is the same underlying grid.

When I create my own graphical objects to use as part of a spreadsheet there might be some flexibility in how to position things. If I am nervous that my input will not be aligned with the present order, not exact enough, I can turn on the option "snap to grid". This is how Microsoft describe it: 'When you draw, resize, or move a shape or other object in the software, you can set it so that it will align or "snap" to the nearest intersection in the grid (even if the grid is not visible) or snap to other shapes or objects' (Microsoft 365 Support n.d.).

The spreadsheet and the grid are some features of digitally engendered ordering practices that have been around for some time in Mundania. They become noticeable as part of graphical user interfaces, but to what extent does the snap-to-grid logic extend beyond the screens?

Snap-to-grid, as it has been utilized in spreadsheet software and other software that is based on grid structures (like music sequencers), is more or less like an automatization of the grid concept in notebooks. Squares and straight lines. The grid is a basic spatial ordering principle that encourages precision and accuracy. A grid notebook made of paper also encourages signs/figures/letters to be ordered and positioned in squares, like a mental hint to think along x- and y-axes. In music software, snap-to-grid is a tool used in relation to quantization. Notes in a sequencer are approximated to the closest discrete point. This creates a well-ordered and predictable rhythmic structure. Notes are played in precise rhythm. In-sync. Such as the repetitive beat of a bass drum.

Snap to grid might be used as a metaphor when thinking more broadly about the technologies of Mundania. About its basics and how it shifts. About computation, and even advanced AI and automated generative systems. The spreadsheet can encourage starting to thinking about quantification and numerical ordering. Grids and algorithms. Scores and units.

Snap-to-grid is about automatic ordering. Sometimes the grid is visible, at other times it is like an underlying force that aligns stuff and processes. Snap-to-grid is a certain kind of optimization. It affords predictability and compatibility. To what extent does it operate at a societal level, to what extent has it been entrenched in new services and habits?

Autosuggest society

Snap-to-grid is a kind of autocorrect. It aligns according to coordinates. It coordinates. Autocorrect is a function more broadly implemented in software that suggests or automatically adjusts what the system recognizes as an erroneous input. It approximates and orders. Such as snap-to-grid. Autocorrect has also for some time been used for automated spell-checking when texts are written in word processors and similar software. If I input the letters *erite*, the software might swiftly turn the letters to the word *write*. This can be defined as a data verification function.

An extension of autocorrect is autosuggest. There is a somewhat blurry boundary between what is a correction or a suggestion. When I write certain phrases in my word processor, I get suggestions, for example to choose more straightforward formulations. It promotes conciseness. If I write 'as long as', I get the suggestion to change it to 'if'. This is not outright correction, but nudging (Gane 2021). I am free to choose, but the system strongly suggests. Recommends? Is there a certain bias hidden here, which nudges people into being more straightforward? No unnecessary words. No unnecessary decorations or ornamentations (see Chapter 2, 'Vanishing Points')? How does this function gradually change the way people write, formulate and communicate? What are the built-in aesthetics, the algorithmic biases? When new norms and styles emerge, they build on something previous.

Autosuggest is related to autocomplete, which has been widely utilized among search engines, such as Google Search (Haider and Sundin 2019: 5). When I entered some signs in Google's search box as it appeared in browsers in 2022, the system suggested a completion of a word or a phrase. It predicted my actions according to certain algorithms. If I put in *smar*, the system suggested search terms such as: *smart wood*, *smartwatch*, *smarteyes*, *smartphoto*, *smart TV*. Since the suggestions were based on popular terms and phrases, the mechanism could further reinforce prevailing differences in popularity of search terms. This was a reinforcement mechanism, a more intricate variety of lists that encourage people to choose what is already highly ranked.

While Google could steer how autocomplete worked, it was also something that could be manipulated by varying stakeholders with different aims. These aims were not necessarily benevolent (Haider and Sundin 2019: 94). Biases of various kinds, sometimes oppressive, became part of the suggestion machinery (Noble 2018).

Suggestions or corrective features can be seen as an optimization of a system, of which the human is a part. The question then is what is considered as optimal by different stakeholders? Here, standards and norms are put in action. Influenced by such things as aims and intentions, by specific designs and the selection of learning data. A gradual shift became apparent during the first half of the 2020s. AI and machine learning was integrated in more and more systems, and new services were offered and promoted. Code, text, images, videos and sounds could be generated and altered with the help of advanced algorithmic systems. Suggestions became more creative, extensive and advanced, enmeshed in conversational interactions. Gradually new services could become normative, while the logic and biases embedded in the output of computational processes became more and more obscure. Humans became ever more entangled in these assemblages of creation, modification and transformation. It is crucial to consider which earlier technologies, preferences and practices the emerging technologies and services were built on. Which stakeholders were involved?

With the emergence of the novelties of the early 2020s, that which had already been mundanized, such as autocorrect and autosuggest, was camouflaged and hidden deeper down the infrastructures. It became harder and harder to discern long-term transformational shifts. In some sense, corrective or suggestive mechanisms are a kind of cybernetic social engineering. An extension of modernist ideas about functional designs that promote certain behaviour, norms and aesthetics, and that uphold order. *Code as Law*, as legal scholar Lawrence Lessig put it when arguing that computerized code and digital systems regulate conduct (1999).

Within the digital cultures of Mundania, norms, styles and behaviour have been influenced or even enforced by software for some time (see also Manovich 2013). But software-influenced creation, alteration and conduct also build on previous phenomena. Earlier pre-computerized systems of standards could steer or promote how actions unfolded. During the 20th century especially, with the strong drive to standardize and modernize society in many countries, several human habitats became widely regulated. Everything from screws to servers is nowadays based on vast systems of standardization. The layers of Mundania are often also layers of standards, upheld by normally disregarded organizing bodies, such as ISO, and consortia, such as W3C. In everyday life, most people do not engage deeply with the workings of standardization bodies or standards organizations. But in a society characterized by pervasive technological systems, people can hardly escape the standards of everyday life.

The evangelics of optimization

Standardized order and autocorrective instruments have been aligned with ideas about good future-proof life and living for some time. A strong focus on the functional and the uncluttered started to gain traction in everything from city planning to interior design during the 19th century, and some decades into the following century the home especially became the place where smooth efficacy should be implemented in countries like Sweden. The rational, clean and uncluttered household became an ideal. Machines for living and effective and clean everyday life were hailed. This was also when the enthusiasm for glass, as discussed in Chapter 6, 'Opacity', took hold.

Modernist literature and culture scholar Victoria Rosner has written about the wishes for domestic effectivity from a North American viewpoint. According to her, changes in the routines of domestic life came into focus especially during the period between the two world wars. The home in itself was approached as a problem that should be solved. Homes should be 're-imagined, streamlined, electrified and generally cleaned up' (Rosner 2020: 2).

In Sweden, initiatives to optimize and streamline domestic life, work and home design also accelerated some decades into the 20th century. Industrial prefabrication of elements, such as kitchen units, made standardization easier. In 1944, HFI (the Institute of Home Research, in Swedish: Hemmens forskningsinstitut) was founded by housewives, and the domestic science teachers' association, together with the public institution Aktiv hushållning (Active Housekeeping) (Brunnström 2018: 88). These organizations were working and striving towards active rationalization of domestic work, and the methods were based on systematic research, time studies and minute observation of domestic practices. Workbenches should have the ergonomically most suitable height, electric appliances should replace manual labour, the layout of the kitchen should provide for smooth movements and work. Not a single unnecessary step should be taken in the well-planned kitchen. It should also be hygienic, electrified, standardized, optimal and efficiently smooth.

The effective home promoted by organizations like HFI was built on certain assumptions and norms. The kitchen was almost like a laboratory, and this architectural style was also sometimes explicitly called 'laboratory kitchens' in Sweden (Jönsson 2017: 76). At this time, many kitchens were built mainly for one person, who was supposed to prepare meals and do the housework. The social norm in Sweden during mid-20th century was also that this person was supposed to be a woman, a housewife. The kitchen should be a smooth and effective place for domestic work, well-planned and organized for the working housewife. Here, ideas about the good and effective boil down to very specific details, such as the standard height of a bench, determined by the average height of Swedish women at the time. The distance between sink and stove or the size of the different kitchen elements was also determined.

The standardized kitchen design from the mid-20th century is a physical device that predates the way digital autocorrect and autosuggest is conceived. A kind of physical nudging and more or less snap-to-grid. Cooking and socializing in a standardized domestic space, as well as screen-based creation and socializing, have different corrective and nudging mechanisms that influence how everyday life unfolds.

Autocorrection merges ideas about the rightful with optimization imaginaries. Autosuggestion brings in ideas about 'own choice', but the aim is still some kind of optimization. When something is optimized it is supposed to also be good. Correct and appropriate. Optimization comes bundled into words like 'smart', as well as words such as 'neat', 'agile', or 'lean'. The latter has been part of an industrial paradigm through which organizations and systems have been geared towards smooth and effective processes. It has been implemented in everything from car production to healthcare. The long-term effects of this management ideal are yet to be seen and scrutinized. Optimization imaginaries are characterized by a strong belief in all-encompassing systems combined with the embracing of the latest technological stunts. This is what technology critic Evgeny Morozov has called solutionism (2013). A kind of tech *hubris*, through which large-scale humanly designed and surveyed systems that are continuously fed with technological novelties are supposed to make the world a better place. This stance could also be called *cyberhubris*, an overconfidence that everything could be part of a controllable system.

A person with a specific device at hand often seems to get overconfident that the specific device, system or technology can be used ubiquitously to solve problems. Hammer and nail, electricity, machine learning and generative AI systems. Snap-to-grid or auto-suggest algorithms. This is how dreams about ubiquitous computing and media take hold. This is how the smart city, the smart home, and new ways and order have been promoted. Often fuelled by *portfolio evangelists*. Representatives for corporations that promote their specific product portfolios as the solutions for present and future challenges. Imaginaries about futures are aligned with what can be offered from specific portfolios. Broad-scale implementations of new technologies are often promoted as remedies against irrational and devious human behaviour. A kind of overconfidence that people and societies with the help of machines can know and control almost everything.

Brandverses

August 2022. Where are my keys? I remember that I had them at home. Or did I accidentally drop them? Did I drop them in the recycling bin when I took out the garbage? What if I had had something attached to them so that I could localize them. Maybe using my smartphone?

Now I have that something. A small thing. It looks like an oversized pill with a round shape and a shiny metallic surface. I can see my face mirrored on its surface. In the middle of my face is an engraved Apple. The small thing is called an airtag. I have fastened it to my keyring, and now I can find my keys using my smartphone or some other device connected to the iCloud system. The airtag has a small radio antenna, and circuits and components that makes it localizable by other Apple products. If I walk by someone on the street who also has an Apple-product, there will be signals between the devices, abstractly saying 'I'm here', while they are virtually 'handshaking'. The airtag has a unique serial number and is associated with my Apple ID. When I attach it to my keyring, my keys become part of a networked geography. Another thing becomes part of the internet. Another mundane but important part of my life has become under the influence of corporate and technological logic. This is how large companies become further enmeshed with the ordinary and instil their logic as an ordering principle of everyday life. This is how *brandverses* grow.

The word 'verse', as in universe, comes from the Latin *'versus'*. It means turning, as in the turning of a plough, to create a furrow. In this sense, a verse is a track or pathway that points out a direction, that orders. 'Universe' means that everything is turned to the same. Is united, uniform, universal. A 'brandverse', then, means that the forces underlying a corporate brand turn things and turn life in a certain direction, they formulate and direct the world. The furrow and way of Apple, Google, Amazon, Microsoft or Alibaba Group. The ways of Mundania in the 21st century are the ways of brandverses.

One commonality among powerful companies is their ambition to build and maintain their own robust and profitable systems where their products and services can thrive. These are brandverses, but within management-speak, other words are used. When it comes to the business of digital cultures, sometimes the systems are referred to as 'ecosystems' or 'platforms', but sometimes also as 'walled gardens'. Even if the latter metaphor was used more seldom in the 2020s.

The economical and strategical paradise for a company such as Amazon, Apple, Google, Microsoft, Meta or Alibaba Group would probably be to be the owners and controllers of a walled garden. A space where they are in control, where customers are kept in an enclosed space, among the products offered in this branded monetary *Hortus Conclusus*. Where corporations turn the soil with their ploughs, creating their lines and grids of furrows and paths. But it is often hard to maintain walled gardens. There is always something outside that might influence and challenge orders. The universe is always larger than the brandverse. And the walled garden is seldom the dream scenario for customers. You are submitted to the decisions, designs, whims and visions of those controlling the garden. The walls might offer some protection, but you can only smell the flowers of this specific garden.

There have been several examples of techno-economic walled gardens. Some successful, some less so. Many game consoles, like Sony Playstation, Microsoft XBox, Nintendo or SEGA, have used walled-garden-like concepts. Among powerful tech-companies, Apple has been famous for controlling compatibility in a way that resembles a walled-garden logic, where hardware, software and services are designed to operate and to be used in closed and controlled systems.

There seems to be a constant oscillation between open and closed systems in digital cultures and economies. The question is what the next move will be. In which direction does the pendulum swing? Will systems be more closed or more open? There is a constant competition between different brandverses and stakeholders, and the power relations might slowly or quickly shift. Sometimes there are stable walls between different interests, sometimes they are permeated. One force that makes brandverses coalesce or interrelate is the establishment of common standards and agreements about order. The internet, as well as several other advanced technological arrangements, is based on a multitude of such standards and protocols. They form, as already noted, some of the infrastructural layers that underpin Mundania.

Logistical media

A way to further understand the system-bound and organized work of technologies and media is to see them as logistical. How are points in space associated with standards and systems, how does verification work? Media are entwined with the development of order, with the ways human operations are instilled, organized and upheld. This is how the turns and furrowing of brandverses coalesce with concrete technologies. Order emerges, becomes elemental, ambient and infrastructural through what can be called logistical media. According to John Durham Peters:

> Logistical media have the job of ordering fundamental terms and units. They add to the leverage exerted by recording media that compress time, and by transmitting media that compress space. The job of logistical media is to organize and orient, to arrange people and property, often into grids. They both coordinate and subordinate, arranging relationships among people and things. (Peters 2015: 37)

In this sense, media instil order and rule-bound processes that humans must align with to be part of worlds with technologies. Logistical media help to steer, nudge or force things into positions. Snap things into imaginary or real grids. Logistical media are, according to Judd Case: 'Lighthouses, clocks, global positioning systems, temples, maps, calendars, telescopes, and highways' (Case 2013: 380). These are ordering devices, based on terms and units. They mediate between the distant and the close. They generate time and temporality and keep it. When logistical media are in place they are normally overlooked, taken-for-granted. They are embedded in the layers of Mundania.

Logistical media influence behaviour. They can, of course, also be altered by different practices and interventions, but they are hard to avoid, and they become crucial for the way societies, people and groups organize. Peters points out further key examples of logistical media, such as: 'names, indexes, maps, lists (like this one), tax rolls, logs, accounts, archives, and the census. Money is surely the master logistical medium – a medium, as Karl Marx complained, that has no content in itself but has the power to arrange everything else around it' (Peters 2015: 37; see also Rossiter 2021). When these media, with all the interconnected technologies, become more and more widespread and pervasive, they become elemental, ambient and mundane.

Addressing

November 2022. In the hallway, there is a hole in the wall. It is poorly covered by a white plastic cover, fastened with two screws. I can see the hole beside the left side of the cover. Through it comes a thin optical cable. It brings the internet connection to the wireless router. The thin cable disappears through the hole, into the wall and it runs inside the structures of the house out to the staircase and down into the basement. It follows the pathways of the old cable canal that was formerly used for the copper wires of the analog landline telephone. The copper phone-cable is since long defunct, out of use.

When I look at the hole in the wall, I remember our last landline phone. It was black and had a winding spiral cord. The metallic sound of the dial tone that came from its small speaker when I lifted it. 440 Hz, the musical pitch of A. Some years ago, I cancelled our old telephone number. It had become redundant in times of mobile phones and the internet. The dial tone that had been constant for decades when the receiver was lifted was silenced, never to appear again through that system, in that place.

When we gave up our landline phone, a phrase became more commonly heard. "Where are you?" This was an often-used phrase to greet a person when calling them on a mobile phone in the first years of the 21st century. Before the mobile phone it was ridiculous to ask where someone was when they answered the phone. The receiver that a person talked into was firmly connected to the telephone via a cord. Like the one that had been in our hallway. The telephone was thereby located at a specific address. The phone number was assigned to a specific physical location. From the phone, a wire and a landline of copper was strung from building to building, through cities and landscapes, connecting different addresses to switches and other parts of the telephone system. The person who answered a phone was at the other end of the wire, at the address where the phone was registered. Mobile phones changed this. People on the move could make and receive calls, turning the idea about static telephone addresses to something outdated. Similarly, when email addresses or electronic mailboxes replaced physical mail as a means of correspondence, the idea about what is and is not an address had to be renegotiated. This was what happened when Swedish Prime Minister Bildt called US President Clinton in 1994 (see Chapter 5, 'Beyond'). New orders emerged. Logistical media orients, directs and evokes orders. It is about rhythms and whereabouts, about tempi and addresses.

The development of IoT basically meant that different artefacts and material objects could be equipped with electronic circuits and components and assigned addresses. IP (Internet Protocol) addresses. If given these addresses, things could be connected via internet and identified as separate units. A thing in the IoT-sense could be anything from a car to an implant, a camera mounted on the top of a mountain, a toaster, a smartphone, or an oven. All assigned a separate IP address.

The verb 'to address' means to direct something towards someone or something, to aim at, to approach. When it comes to the noun 'address', it has multiple variations, especially as it is coupled to different practices, structures, and technologies. There are street addresses, email addresses and IP addresses. They all make logistics possible, and when addresses shift character logistics also change. With IP addresses or email addresses, an address does not have to be stationary, not assigned to a specific geographical place, as had long been the case with addresses. The IP address does not have to tell anything about geographical whereabouts. For geolocation of connected things, other systems would have to be employed.

The idea about fixed addresses can probably be coupled to the rise of early human settlements, and maybe also to graves and burying grounds, as John Durham Peters put it, referring to Lewis Mumford: "Settlement always goes with graves; Mumford suggests the dead may have been the first humans to have a fixed address." (Peters 2015: 145). This kind of address is something other than the mobile IP address. It is the final (physical) destination, the end point, the terminus. The dead might be associated with long-term fixed addresses, but living people are also often associated with specific addresses. The idea about persons living at settled addresses has in many countries for a long time been coupled to population registration and citizenship. It has been associated with ownership or rental of real estate and with wide-spanning systems like postal services or utility services. In Sweden, a permanent address is part of the population registration, administered by the tax authorities. The definition of this address is the place where a person normally sleeps. In the Swedish law for population registration (Folkbokföringslag (1991: 481), it is the place for daily rest '*dygnsvila*'. In a sense, this kind of address is also defined as a resting place, albeit not a final one.

In Mundania there is a layering of addresses. In Sweden, a permanent physical address is the norm when a person applies to receive services from a bank such as a bank account and a bank card. Once the person gets the bank card, the plastic sheet can be used to purchase digital connectivity, to use devices that have assigned IP addresses, to get access to online services and to create email addresses. Using these layered systems of addresses, a person could order something that would be physically delivered to a specific address. The question is what happens if some of these addresses become inaccessible?

People can be associated with addresses. They can also be associated with certain things. Various kinds of documents have been used for identification for centuries throughout the world. Documents such as passports have been part of the emergence of what media historian Craig Robertson has called 'a documentary regime of verification' (2009). Documents, such as the passport and, later, various kinds of plastic identification cards, have existed alongside and enmeshed with numerous registers of people's identities and personal data. Registers not just managed by official bodies, but also often initiated and owned by commercial actors, such as insurance companies or credit rating agencies (Husz 2022).

When it comes to addresses, identification, authentication, and communication, a gradual shift is taking place. The fixed address of a person has for a long time been a kind of entrance point for logistical media in Sweden. It is through fixed addresses that persons get access to systems (like welfare and utility services). Will this change if the constant tracking of people's physical movement becomes the norm? The question "where are you?" when someone calls by phone is about to become redundant again, when more and more digital services disclose the whereabouts of people. Hiding your location for certain systems, will maybe be as norm-breaking as not having a permanent address during former decades. To be constantly tracked and monitored will most likely be the deal in many contexts. What could it mean to never be able to hide your whereabouts? For whom, for what? If this would happen, it would not be domestication of technologies, it would be mundanization. A gradual transformation and acceptance of increasingly inexplicable circumstances.

The emerging ordinary

If addresses are meant to work within a system, they must be compatible with the system. They must follow certain standards and protocols. It is not a recommendation or norm, but a prerequisite. If you have a house without an address registered by the postal service, no packages will arrive. You will be external, off the grid. Beyond the reach of systems. What if a product supposed to be networked were not to have a valid IP address?

What will the future addresses of Mundania be? The future protocols? How will future logistical media be comprehended and imagined? When posing these questions, it is important to stress that there is never any unison imaginary or experience when it comes to such seemingly straightforward phenomena as addresses or standards. No clear once-and-for-all definition or order. Mundania seems to be comprehensively homogenous, but it varies depending on position, situation and circumstances. Physical addresses have very different meanings for people, from the homesteader to the lodger, from the owner of palaces and mansions to the paperless migrant. Similarly, new variants of addresses will be imagined, dealt with and experienced differently. Addresses are also to a large extent something that is beyond humans. What new logistical media make evident is that addresses do not have so much to do with immediate relations between people. When more and more things get IP addresses, most addresses are not directly associated with humans; instead, addresses are about devices communicating with devices, sending, receiving, processing, synchronizing, verifying. Handshaking. Addressing.

The role of logistical media based on coordination and subordination is crucial in Mundania. Categorizations, procedures and temporalities are often experienced as straightforward. But they are anchored in incomprehensible technological assemblages, they develop over time and meet the variabilities of everyday life.

Some categorization principles are based on corporate logic and are manifested through brandverses. Corporate logic, together with various systems and standards, steers what is possible and not. What can be addressed? What is compatible and possible to use and combine? Much of these logistical workings are ignored once they are established. They become almost like the air people breathe. All around yet ignored. In-between, beyond, and beneath. Technological systems, strategies and labour spawn and uphold processes through which the ordinary keeps on emerging in Mundania.

8

Variability

At a closer look, Mundania is constantly re-created. Order is, as we have seen, merely temporarily and situationally fixed. Mundania is like an image on a computer screen that seems to be static, but beneath and beyond the glass surface of the screen there is a constant flow of energy, processes and invisible work upholding what is experienced as a motionless image. Mundania keeps on emerging as people experience it.

What is experienced as ordinary and what is experienced as inertia or transformation is contextual. The different experiences of change, interferences and suspensions are furthermore unevenly distributed. There is variability and mutability in Mundania. Conditions might change and differ between time and context. Shifting tempi, temporalities and horizons of possibility for different people. This variability is related to social processes and stratifications and to the ways imaginaries unfold. What is possible, how does variability emerge and how is it maintained? The here and now of Mundania is influenced by projections into different times and possibilities, by expectations of what could come and what is remembered.

Elsewayness

How to capture the fleeting now, the emergent contemporary, and all the complex processes that continue, emerge and enmesh throughout time? How to capture all the varied rhythms and processes that uphold Mundania? How is the very local connected to distant places and to larger and stronger ongoing changes? Futures to pasts? All the incalculable potential entanglements. Uploading of files to a cloud service and climate change. Configuration of a webcam and the fluctuations of stock values of tech corporations on Nasdaq, the cosy hue from the light of a smart lamp and the death and suffering radiating from conflicts over rare earth metals? What is linked to what, and what is relevant to consider?

How the future of Mundania is manifested is not predetermined. It is important not to fall into the trap of technological determinism, to think that the specificities of the technologies that we are enmeshed with will inevitably steer us into one specific future (Rahm and Kaun 2022: 24). The future is inherently uncertain, unknowable and enigmatic, and attempts to steer or predict what will happen often fail. This is why it is relevant to think about futures in plural (Pink 2022: 14-15).

It is important not to tell that linear and straightforward story about inevitable technological progress or peril once again (Jasanoff 2021: 16). Yet, technologies are imperative. They do not predetermine, but they influence and nudge into position, and they autocorrect and suggest according to specific designs, logics and criteria. Futures are shaped by practices, influenced by what technologies make possible. In this sense technologies are orientational, but they do not predetermine and inevitably steer us in one specific direction. A strong influential force comes from that which is inherently intangible, from imaginaries. Fleeting and often variable ideas about coming futures. Visions that spark missions. These can turn to something very concrete. Ideas and imaginaries can be a remedy, or they can set the world on fire.

Imaginaries, albeit fuzzy, have the potential to crossbreed the contemporary with alternatives, with alterity, with that which could be different. Here and now is amalgamated with elsewhereness, with that which is beyond. It is also amalgamated with projections of what might become or should not become. Imaginaries can evoke what could be called *elsewayness*. That which could be fundamentally different. In this sense, imaginaries have a transformative potential. But drastic elsewayness does not seem to be the modus operandi. Imaginaries seem to be hard to shift in radical ways. Imaginaries are indeterminate, but patterns recur. Just like there are path dependencies in the development of technological systems, there are certain preconditions for the ways imaginaries often develop. Imaginaries seldom reach escape velocity. They often gravitate towards certain larger prevailing constructs that determine their (fuzzy) arrangement and orientation. It is, for example, hard to think about a world without money or financial systems like the ones prevailing today. To conceptualize societies and the world as some variation of capitalism is a major gravitational force from which imaginaries can hardly escape today. It is also hard to envision societies without humans entering certain roles, careers and professions. It is hard to think about renewal and improvement without drifting towards established tropes, such as innovation models or evolution. Even ideas about radical innovation and revolution are rule-bound.

Imaginaries can be oriented towards what could be different, towards elsewayness, but the elseway is often tinted by preconceptions, and by biases and specific interests. The faint hum permeating Mundania whispers of inevitability and inescapability. Annette Markham discusses a similar dilemma in an article about what is possible to imagine when it comes to technologies.

Any telling of ourselves, individually or culturally, is part of a pre-existing narrative environment, as Judith Butler's extensive work on the performativity of gender illustrates so poignantly. Thus, anything we might call an imaginary is in some ways stuck within (or departing from) a predefined template for context, content, and genre, and sometimes all three. (Markham 2020: 385).

It means that the way humans imagine worlds is determined by that which Markham calls predefined templates (Markham 2020: 385). By discursive frames. These very frames are often invisible until they are challenged or destabilized. These templates, together with routines and procedural and recurring behaviour, keep some things as they are. Generate continuity. Despite a lot of talk about disruption and transformation, much stays the same. The rhythm of Mundania is this peculiar mix of change and repetition. Some conditions change, while others remain. That faint hum is still here.

Another 1984

October 2022. Ding! Another email arrives. This time it is a brief notification. It tells me that an amount has been withdrawn from my credit card. This specific withdrawal happens once a month. Every time I feel a slight relief that it works. Seamlessly. Simultaneously, I feel a slight annoyance. A feeling of dependence. This withdrawal gives me access to cloud storage for one more month. It means that files, films, photos, works and apps are stored somewhere in a distant data centre. What if I would like to change this deal? How would I end this subscription? How should I start moving files and start to deal with all that which has taken place beyond my attention for some while? Which bindings are connected to this cloud account? Which relations? What are the implications to end this? Often one deal would affect several others. The thought provokes a slight vertigo. It is not impossible, but quite troublesome to leave a brandverse. Sometimes I feel like a mindless drone that just keep on paying to get access to some labyrinthine relationscape that I can neither escape nor live without. How did I get here? Gradually I guess, step by step, file by file, click by click, nudge by nudge.

Through the years a kind of tentacular embrace has taken form. I have confirmed agreements that I have not read in detail. I have confirmed that I agree to terms and conditions by restlessly clicking when I have started to use some service. Clickwrap agreements are all over Mundania. To the technological and economical ungraspability of Mundania, we can add the legal. These circumstances started to build up some time ago.

A certain year that is often evoked when it comes to matters of control, dominance and systems hard to escape is 1984. At that time, Apple aired a television commercial for their Macintosh-computers that has become iconic in advertising circles. It appeared on television through several outlets, on one occasion in a break during the Super Bowl, the final of the US National Football League (NFL). The commercial was partly aimed towards new creatives who wanted alternatives to the business machines provided by the computer giant of those days, IBM. In the short commercial directed by Ridley Scott a theme from George Orwell's novel *Nineteen Eighty-Four* was taken as the point of departure to convey the message that Apple's small 'more human' computers should break the regime of the prevailing computer industry. The plot takes place in a large industry-like environment, in drab and uninviting corridors and halls. Grey colour in the corridor, and when entering the hall, a cold bluish hue. IBM was called 'Big Blue', and the corporation was hinted at as a kind of Big Brother in the commercial. Rows of drone-like bald and uniformed people march in rows and sit down in front of a gigantic screen (Wikipedia, 1984 (advertisement) n.d.)

In this blue-grey bleak environment, something pops out. A blonde-haired woman in athletic clothing comes running. She wears a white top and bright red shorts. Her colours are graded differently than the rest of the footage. She looks bright and colourful against all the drab darkness. In her hands, a huge sledgehammer. She is hunted by threatening men in protective gear. They appear as militia or law enforcement officers from the dictatorship of some dystopic science-fiction film.

The sound of the commercial is based on heavy boots against the floor, a siren or horn that repeatedly sounds and reverberates in the hall. On top of that a male voice. It has an authoritative and commanding tone. The voice of a leader, possibly Big Brother, who also appears on the huge screen. The voice announces that:

> Today, we celebrate the first glorious anniversary of the Information Purification Directives. We have created, for the first time in all history, a garden of pure ideology – where each worker may bloom, secure from the pests purveying contradictory thoughts. Our Unification of Thoughts is more powerful a weapon than any fleet or army on earth. We are one people, with one will, one resolve, one cause. Our enemies shall talk themselves to death, and we will bury them with their own confusion. We shall prevail! (Quoted in Wikipedia, 1984 [advertisement] n.d.)

Before the voice shouts "We shall prevail!", the woman spins round and hurls the hammer. Like an athlete from the Olympic Games. The hammer hits the screen with an explosion. White light spreads through the room. The rows of people stare, shocked, with open mouths at the disruptive action. Then comes Apple's message in text and voice-over: "On January 24th, Apple Computer will introduce Macintosh. And you'll see why 1984 won't be like '1984'" (Wikipedia, 1984 (advertisement) n.d.). The commercial ends with the Apple logo, which at the time was rainbow coloured.

This was around forty years ago. The ideas about tech-based creativity, as promoted in the commercial, have to some extent taken over much of computer business and the way consumer electronics are understood. Now there are new giants reigning. One of these is Apple. The underdog has grown into a powerful position. Sporting a more cheerful and laidback appearance than earlier providers of business machines, but yet very powerful and influential. There are also new stakeholders, roles and business dynamics. Much has changed since 1984, but much has also remained. Revolution and disruption, as it was evoked by Apple in 1984 for example, has become routinized, almost fetishized within Big Tech and digital cultures.

Cheerful colourings and cartoonish appearances have become the main tenor of several of the prevailing brandverses. Even if Apple has changed its rainbow-coloured logo to a sleeker monochrome appearance, the colourfulness appears in many other places of their brandverse. Power and trustfulness are often conveyed differently in the 2020s than forty years earlier. One of the world's most powerful corporations, Google, can have a brand identity colour-coded more like a preschool than earlier global business conglomerates. But the question is what remains, under the cheerful colours and playful designs. How would you escape the entanglements of conglomerates and brandverses and the rhythms they instil?

Acceleration imaginaries

One persistent template of imaginaries that has been around in Mundania for a long time is about rhythm and tempo. It is about increased speed. Over the past decades computing power has undoubtedly increased. The technological possibilities of a computer have multiplied compared to those that were promoted in 1984. The increase in computing power has, together with broader technological development and certain economic circumstances, fostered imaginaries about perpetual acceleration.

The increase in computing power and acceleration has often been associated with what has been called Moore's Law. In 1965, the American engineer Gordon Moore saw that the number of transistors in a silicon chip had doubled every two years. For a special issue of the journal *Electronics*, Moore made a projection into the future. He expected that this trend would continue, and that the power of computer components would increase exponentially, while simultaneously decreasing in relative cost. Since then, this trend has been approached more or less as an inexorable truth, almost comparable to laws of nature as formulated by Newton or Einstein even if there has also been debates about when and how the trend would end (Britannica, Moore's law, 2023; see also Rotman 2020).

Computing power has undoubtedly increased over the years, and Moore's law has been influential in tech-based businesses, visions and imaginaries. But how far can this logic be stretched? It is like the dreams about speed and fast, streamlined mobility that emerged during early 20th-century modernity and in Modernist art have been ingrained in imaginaries and repeatedly come back in space races, cyberspaces and computer-driven AI-visions. Are the imaginaries about increased speed and computing power related to unspoken ideas about escape velocity, about leaving the old world and the bads of the past behind by embracing acceleration?

Moore's law and its associated logics have been associated with inevitability, a march of technological progress and a kind of persistent upgrading ethos (Willim 2003a). It says that upgrades will continuously come, that they will provide us with new opportunities and that people better align with this pace to 'not be left behind'. Trains, planes and AI. An upgrading ethos says that technologies will continue to increase in power and their benefits will spread over the world, leading to Society 1.1, followed by 1.2, 1.3, ... 2.0 and so on. This is one of the most persistent templates of Mundania. The imaginary about upgrading. Based on ideas that new versions and things will come and that they will foster transformation. A promise epitomizes the upgrading ethos. It says that even if present things are far from perfect, they will be fixed in a coming upgrade. Rest assured, the upgrade will solve things.

Upgrading can also be associated with a physical thing getting new features, maybe by the installation of electronic components and software. Like an upgrade of a lawnmower or an oven. The newly upgraded thing gets a new quality, it becomes electronic, digital or connected. Often this upgrade has been promoted by the word 'smart'. Smart toothbrushes, hairdryers, lamps and materials. A couple of these were discussed in Chapter 3, 'In-between'.

In this sense, an upgrade can be a qualitative or quantitative enhancement, but foremost it is an incremental change that improves what is already there. New and tweaked functions or more capacity, like more memory or bandwidth. An upgrade is same, but different. An accumulation or enhancement on a fixed scale. It is based on the power of addition and increase, the power of plus. The idea about incremental improvement has been a crucial aspect of Mundania. It has to do with the development of new electronics components such as microchips, but it is foremost associated with the development of software. It has become a given that, regularly, new versions of software will be developed and introduced.

Upgrading does not always come smoothly. It is in practice based on rhythms that are not always in sync. Which software versions work with specific hardware? New processor architectures or other changes of hardware often require that software is rewritten. The releases of hardware components and compatible software does not always coincide. There might be lags and different tempi leading to compatibility issues. Neither do new generations of software support hardware that has reached a certain age. When new software platforms or major upgrades of operative systems are released, earlier versions of third-party software might also stop working. This is more a rule than an exception. Then it is worthwhile to wait. A rule of thumb is that the more components from different producers and brandverses that are part of a system, the longer the wait before it is safe to install a new system upgrade. One of the competencies of Mundania has been to get hold of these rhythms of compatibility. To believe in coming better versions, but to also know when to wait before installing. When can and should I upgrade software? What happens with my order and routines and my software when I change hardware? What is compatible with what? Upgrades are supposed to make experiences smoother and smarter, while the process of upgrading per se is a more intricate matter.

Temporal alignment and getting hold of different rhythms are an important part of mundanization. The cyclical shift between day and night and the yearly cycle of seasons are accompanied by such phenomena as upgrades as well as update cycles, subscription periods and charging times. There are variations between all these different rhythms. One specific difference is the one between upgrades and updates. When it comes to most software, upgrades are more drastic and obvious changes, while updates are more frequent fixes of the code. The latter often include security patches and removal of glitches and bugs, and this becomes more of a requirement. Users are supposed to update and to stay in sync. Just as they are also supposed to back up data regularly. These practices are from security angles, referred to as Cyber Hygiene or Digital Hygiene (Dharampal 2021). Moral demands to be 'hygienic', to act responsibly and to align with the rhythms of updates, security patching and the recurrent management of passwords and file systems give rise to ideas about the good compliant digital citizen (Dharampal 2021). The good digital citizen keeps products alive by updating them. Software, or objects that are not updated are dead, as Wendy H.K. Chun has put it. 'Things and people not updating are things and people lost or in distress, for users have become creatures of the update. To be is to be updated: to update and be subjected to the update' (Chun 2016: 2).

In Mundania you are supposed to update to stay in sync and to be 'hygienic'. To be alive? This can be compared to how citizens of Sweden in the mid-20th century were nudged into position by specific designs of homes and everyday environments (as described in Chapter 7, 'Order'). The question is, which stakeholders are involved in these different corrective agendas?

People are nudged into certain rhythms. And it has become increasingly easy to keep up with the pace of updates. During the first years of the 2020s, updating has become more of an automated process. People are asked if they want their software to update 'in the background'. Once they agree to this, updates become part of continuous ambient improvement. It removes conscious practices of updating. This is a further strengthening of tech-infused Mundania. Improvements happen in the background. Variations and transformations take place beyond awareness of the person using the system. You automatically stay in sync, quantized. This ambient improvement must however be foolproof when it comes to compatibility. If it should break some of the functionalities between installed software and connected hardware, the convenience of ambient improvement could instead lead to major concerns. Despite this, the trend towards more and more automated processes was strong. Several processes became ambient in the early 2020s.

Conceptual congruity

Imaginaries do not take form in a vacuum. Tasks and thought are entwined. Imaginaries often emerge, and become strengthened when people engage with, think about and with, and feel for things, technologies and systems. 'There's a general argument in the history of technology that we use the dominant technology of the day to think about the world arounds [sic] — the brain was hydraulic (think Freud) when dams dominated; a switchboard when there were telephones abounding and is now a computer' (Bowker 2020: 96). This is how Geoffrey Bowker put it when he discusses how imaginaries around archives have developed and mutated. Technologies and imaginaries merge. Cybernetics and electrical circuitry merge with imaginaries about mental energy, with bodies and mind to be charged like batteries. Drained and supercharged. Neural computing and GAN (Generative Adversarial Networks) merge with organizational competition and contest as generative processes of quality enhancement. Technological concepts amalgamate with the ways in which people imagine and conceive the world.

When Lev Manovich in 2001 wrote about 'The Language of New Media' he stressed the importance of conceptual transfer, how concepts and ideas from computers would transfer to social and cultural patterns. He proposed that 'cultural categories and concepts are substituted, on the level of meaning and/or language, by new ones that derive from the computer's ontology, epistemology, and pragmatics. New media thus acts as a forerunner of this more general process of cultural reconceptualization' (Manovich 2001: 47). This standpoint harmonizes with the way Bowker discussed imaginaries in the previous quote . It is also exemplified by the way in which upgrading has become a broader concept, beyond software development and use.

Conceptual transfers are hard to avoid. In Manovich's effort to extend media theory to encompass also computer-generated work, he himself contributes to the transfer when he takes concepts from computer science to use in media theory (Manovich 2001: 48). It seems to be a recurring issue, how prevailing technologies seep into the way humans, individually and collectively, imagine the world. We use such words as 'shift gear' and 'hack' to describe practices beyond the engagements with technologies. Also in this book these reconceptualizations occur quite frequently. The black-box metaphor is a good example of how a technological concept seeps into more and more contexts and also influences the way the world is imagined. It becomes a template; it directs associations and make certain thoughts and imaginaries possible. The black-box metaphor in itself paradoxically frames, steers and encloses the ways in which we think. As noted in Chapter 7, 'Opacities'.

When at the turn of the Millennium I was studying the rise and fall of the Swedish internet consultancy company Framfab (an abbreviation for the Swedish name that translates as The Future Factory), I tweaked (yes, a word related to the practices of mechanics that is often used in tech-contexts and beyond) the ideas about conceptual transfer to instead use *conceptual congruity* (2003a). I proposed that a kind of mimetic mirror play between technological concepts and processes and other related phenomena took place. At Framfab, for example, they used the toy LEGO as part of their promotion and as a playful prop in their offices. The qualities of LEGO, its colourfulness, its associations with play, as well as its modular affordances, were mirrored both in the organization (Framfab's organization was characterized by playfulness as well as a modular structure of smaller offices) as well as in some of the digital products and offerings by the company (like the product Brikks) (see also Coupland 1995; Willim 2003b).

Conceptual congruity is something that reinforces specific imaginaries in certain times and contexts, not just through conceptual transfer from technology to the surrounding world, but in a more intricate play between concepts, technologies, aesthetics and practices. This is how clouds and mist merge with the industrial, with grids and algorithms. How (technological) upgrading and (economic) value increase merge. How ideas about machines getting human features, such as smartness and intelligence (or also having hallucinations) comingle with the ways humans should be orderly, professional, uniform, follow procedures and protocols. Things, stories, concepts and processes seem to just 'fit'. Simultaneously creating both congruities and variation. They seem to be part of an unwritten but noticeable meaning-making scheme or pattern. Congruity. Correspondence, or Resonance. The latter word, 'resonance', is often used in electronics and in relation to sound. In Chapter 2, 'Vanishing Points', I discussed it in relation to the tweaking of rotary knobs. Anthropologist Susan Lepselter uses resonance to discuss how collective narratives and poetics emerge and are maintained, and how these can also be influenced by the uncanny, by the power of the unseen, by affect and *apophenia* (Lepselter 2016). 'Resonance is not an exact reiteration. Rather it's something that strikes a chord, that inexplicably rings true, a sound whose notes are prolonged. It is just-glimpsed connections and hidden structures that are felt to shimmer below the surface of things' (Lepselter 2016: 4). Resonance can be something that excites and enforces, that feels right, like conceptual congruity. But it might also have a darker tenor, evoking the uncanny. Resonance and conceptual congruity can also be strengthened by atmospheric attunement and ecstatic things, as discussed in Chapter 3, 'In-between'. The ambience, the atmosphere, and the ways we engage with things are entwined with the ways in which imaginaries unfold and varies. How they resonate with the world.

Citius, altius, fortius

To further put the focus on conceptual congruity, let us once again think about the static screen image that introduced this chapter. It is maintained through fluid and rapid processes that go unnoticed by the onlooker. Let us use it as a metaphorical vehicle. Under the surface of the screen and in its connected peripherals, quick variation and alternation keep things in their places. Electricity and processes in and between circuits. The experience of the static image on the screen is not problematized until some process seizes, is interrupted, or slowed down. The emergence of a flicker or unwanted change of the static image might mean that something beneath the surface has stopped working.

These circumstances might be good to think through. The speedy processes that uphold something seemingly static. When is speed and variability experienced and when can it even be experienced as stasis? In an advanced system, with many components and people involved, a lot of running must be done to keep things in place. But who or what does the running and where does it take place? A thought experiment in Mundania might be to ponder on what processes and labour are required to keep conditions as they are. When and where are increases or decreases in speed or workload noticed and how are they experienced? By whom?

The subheading for this section of the chapter is Latin, it means Faster, Higher, Stronger, and has been the basic motto of the Olympic Games for around a century. Anthropologist Thomas Hylland Eriksen writes about the motto and how sports, as well as today's prevailing logic of business and technology, are based on a conundrum of competition that can be referred to as 'The Red Queen Effect' (Eriksen 2021). To what extent does this effect characterize Mundania? The Red Queen appears in Lewis Carroll's novel *Through The Looking Glass*. In the fictitious and strange realm of the book, a race takes place between Alice and the Red Queen, a race in which Alice must constantly run faster to stay in the same place. When Alice, exhausted after the run, says that in her world you used to get to somewhere when you ran very fast for a long time, the Red Queen answers that it sounds like a slow country, and that here it takes all the running you can do to stay in the same place. This fictitious race has been used to describe everything from evolutionary biology to competition in economics. This paradoxical logic of competition is sometimes referred to as 'The Treadmill Paradox'.

> Although it may look like an accelerated standstill, treadmill competition drives evolution, forcing species and individuals to improve their achievements relative to others over the generations. The 'competitive edge' often invoked in technology and business refers to a quality enabling a company, product or activity to 'edge' ahead of the others, who will in turn have to follow suit. There is a resemblance, in other words, between the cheetah evolving greater speed to catch gazelles who are nimbler and faster than their ancestors, and mobile phone developers looking to eclipse their competitors with a sleeker design, better camera or larger screen. (Eriksen 2021: 2)

Eriksen relates this phenomenon not only to the ethos of continuous improvement and acceleration that has characterized modernist ideas about progression, but also to several societal areas in which competition is intrinsic, such as sports and capitalism. Just like the computers from some years ago are slower and less powerful than the ones launched today, athletes are supposed to run faster, jump higher and be stronger than their predecessors. Citius, Altius, Fortius. Olympic Games and Moore's law. There seems to be a conceptual congruity between several societal fields. A kind of Champions League tenet from the worlds of sports resembles the way businesses and technological innovation are supposed to develop. New resources and innovation of technologies and techniques are coupled to quite a frail idea about fair play, an idea that is challenged repeatedly when the outside world seeps into playing fields and markets. What is fair and what is sustainable? What feeds acceleration imaginaries and the Red Queen race? Where are resources unearthed? Oil money that boosts football teams and venture capital that is pumped into businesses is often based on extraction and gains from unsustainable practices. This is the backside of the acceleration ethos that takes place in the limelight of stadiums, on radiant screens as well as in the most widespread systems and services of Mundania. All the labour and running taking place somewhere else to keep the image in place.

Curve surfing

Another template of imaginaries that has prevailed for decades is that innovation is fostered by disruption. To some extent it stands in stark contrast to the incremental improvements of upgrading. The innovation-through-disruption-imaginary is based on ideas that products that are more fit than earlier ones will replace what was before. Often in ferocious ways. That was why Facebook for a while could have 'Move Fast and Break Things' as their slogan. The ideas are often underpinned by the economic theory about *creative destruction*, as it was introduced by economist Joseph Schumpeter in 1942 (Schumpeter 1994). Creative destruction has been a kind of ethos among tech entrepreneurs, and businesses for decades. According to this logic, the future becomes a better place through entrepreneurial competition and disruption.

What does happen when something old is disrupted and something new is about to be introduced? How do technologies and novelties spread and gradually or abruptly change people's lives? There are several imaginaries about technological innovation that prevail. These can often be coupled to specific temporal patterns, and they are often visually represented by curves and charts. Mundania is full of *curve surfing*, of imaginary rides along curves that show transformations and variations. The progress of technological innovation is imaged and imagined through curves. So are stock market fluctuations, pandemic contagion and so on. These rides are imaginary spatial moves across a grid, along *x*- and *y*-axes in a two-dimensional field. Curves show progress and temporal variation, and they have become hugely mundane and are therefore often taken-for-granted. What does all this curve surfing do to uphold certain imaginaries about progress, fluctuations, trends?

A recurring format for curve surfing has been the S-curve. It has for some time been used to show how innovations are spread. I have seen it in old ethnology books that showed the geographical spread of new variants of rakes and other tools in the Swedish countryside (Bringéus 1976/1986). But I have foremost seen it in numerous forms when it comes to emerging technologies. The S-curve has also been outfitted with phases and stages that should evoke technological progress and adoption. A good example is the *Gartner Hype Cycle*, provided and promoted by consultancy firm Gartner. It is a model, a graphic representation that is supposed to show how rumours appear around a coming technology and how the technology then (eventually) become popularized and pervasive (Gartner Hype Cycle n.d.). According to Gartner, their clients use 'Hype Cycles to get educated about the promise of an emerging technology within the context of their industry and individual appetite for risk' (Gartner Hype Cycle n.d.).

In the first phases of the S-curve, the line stretches out almost horizontally, to then climb exponentially, and to then once again become almost horizontal. At the very beginning of the first part of the curve there is a lot of hype about 'the next thing'. At this phase, there is yet no concrete product or offering available. The possibly coming thing is what is sometimes called *vapourware*, alluringly noticeable in a hazy future (Atkinson 2013; Willim 2017a). Then the curve starts to rise slightly. This is when there are quite few people who have adopted the novelty. Then comes the exponential climb, where the thing becomes really popular. This is followed by the last almost horizontal phase, when there are relatively few new people who adopt what is no longer broadly considered to be a novelty. There are several varieties to the uses of the S-curve, but many utilize ideas about the anticipative power of hype. The curves and the imaginaries about hype become templates that direct associations. These models have also become something of a vernacular imaginary surrounding technologies. They are built around the 'performative force of expectations' (van Lente et al 2013: 1615). The logic of the S-curve is underlying how technologies are imagined in Mundania.

Adoption and adaptation

The first part of the S-curve can be associated with ideas about 'being first' or early adoption. For many, early adoption is something remote, but in a society in which ideas about progress through innovation are prevalent, early adoption gets a certain meaning. Sweden is this kind of society. Of course, it all depends on what is adopted, but early adopters are sometimes considered the heroes of innovation economies, the avant-garde, the pioneers, the progressive, the trailblazers. The ones that help bold innovators, entrepreneurs and start-ups in their imagined pioneer work. Often, technological adoption is a consumption endeavour. Therefore, early adopters are sometimes called *Lighthouse Customers* or even *Alpha Consumers* (Wikipedia, early adopter, n.d.). They are, according to this logic, brave enough to embrace the new, to be ahead of the curve, to jump the train early, to be early onboard the ride to the future. They are not just in fashion; they are supposed to be ahead of their time (Klinkmann 2005; Löfgren 2005). Role models. Influencers. They take some risk by engaging with beta-versions and early versions of technologies and products under development. They are hailed in the rhetoric about innovation.

The early-adopter concept comes from Everett Rogers' ideas and book about *Diffusion of Innovations* (1962). How, why and at what rate do innovations spread? Rogers' model includes innovators, early adopters, early majority, late majority and laggards. It describes how early adoption turns to mainstream adoption, then to a later phase when 'laggards' or late adopters start to, maybe reluctantly, allow an innovation to be part of their lives. One point of the curve has become somewhat like the holy grail of prediction industries, advisors and consultancies working with prognoses and forecasts. It is where the curve starts to rise, the point between early adoption and majority use, when something reaches critical mass. To know when this happens can be valuable knowledge. When does something start to become successful, what will become 'the next new thing', and when?

Rogers' model and ideas, like many predictive models, do not expound that much on social and cultural complexity and various irregularities that can be related to technological change. It is based on a simple and straightforward scheme, which has also made it very persistent and widely used to describe and maybe forecast the processes around emerging technologies. The question is what practices and imaginaries this kind of model brings with it. Does it feed acceleration imaginaries, and the Red Queen effect or Treadmill paradox?

How do the ideas about the virtues of speed and early adoption influence how Mundania is experienced and imagined? What does it mean to adopt something? Adoption is not that far from adaptation. Both words point at a transformative process. Some new element is introduced in a context, and it changes what was there before. The change can be slow and barely unnoticeable, or it can be dramatic. New conditions, opportunities, threats and, after a while, also new habits and routines. The early adopters are ready to take on this transformation, to adopt a service or device and to also be adapted to a system. Early adoption is often promoted as a proactive manoeuvre, an acquisition or takeover. But it is often disregarded that it is also a submissive action. Adoption is, to some extent, adaptation, a defiance, acceptance and submission to a system or technology. When this process is seen as almost a law, a certain pattern of inevitability is reproduced and contributes to a reinforcement of mundanization processes.

Patterns of adoption are central templates when emerging technologies are imagined in the 2020s. These patterns reproduce narratives about progress and revolutions as well as 'origin stories' (Mattern 2017: xxviii). Imaginaries about innovation and how the new emerges have become routine and ordinary. Variations on the same theme. When these imaginaries turn to models, they frame what to expect, how futures are evoked, and which processes are taken-for-granted. The reproduction of these imaginaries about time and technologies seems to stabilize Mundania.

Repetition

Circumstances have definitely changed during the last decades. New technological possibilities and new normals. New devices and greater infrastructural depth. The buzz about the novel has, however, repeatedly turned into a faint background hum. Hypes have emerged and receded. The temptation of the novel has recurrently appeared and faded away. Things have withdrawn or disappeared. Become part of the infrastructures or the media environment or, for that matter, faded into oblivion in the graveyard of failed products.

In Mundania, innovation imaginaries might stabilize. A path dependency of the mind arises. There can be something comforting when the expected rhythms of change keep on repeating. Despite the hyperbole and the promises that a thing, an innovation, or technological achievement will be something completely new, it often feels quite familiar. Hypes and curves. A new supposedly revolutionary thing is promoted, again. It feels like 1984, when the Mac was supposed to shape the future and ensure that 1984 did not become like the version evoked by George Orwell. Or it feels like the 1990s, when Microsoft was supposed to start things up with Windows 95 and their new Start Menu. Or the shift of the Millennium, with all the promises of escape velocity into cyberspace. Or like the time of Web 2.0, some years later. Revolutions, speedy shifts, and disruption have been repeatedly hailed, and sparking novelties put in the limelight. All this recurrent buzz about change and innovation can, in itself, feel numbing and almost static. Maybe comfortingly predictable, even boring.

In this sense, Mundania seems to be stable. Like a static image on the screen. The treadmill spins, things get started and terminated, while nothing much seems to happen. Upgrading processes are going on, according to quite a predictable pattern. Then, buzz, some disruption gains attention. Again. Followed by a period of upgrading. Until another buzz sounds, another disruption occurs, followed by adaptation and upgrading. Then it is all repeated. Routinized revolution. A new normal keep on emerging.

Meanwhile conditions also change. Cycle by cycle. Gradually, while people are busy doing something else. It might be in the accumulation of incremental changes that the radical lies. The infrastructural depth become greater, layers crop up, the role of logistical media becomes different, the atmosphere technologically thicker. The beyond comes closer and closer, and ever more intimate. Mundanization takes further hold. Fundamental change happens, but gradually. While people are repeatedly captivated by the loud buzz of incoming novelties, the faint background hum of the ordinary gradually shifts.

9

Openings

We have reached an endpoint. This last chapter is a kind of terminus, but not a boundary. It opens rather than closes the book. It resonates with the truism that every end is a beginning. Therefore, several contrasting considerations will be presented and juxtaposed in an exposition of short stories; some quite concerning, others slightly banal. These can be understood as openings. Here be peculiarities, failures and frictions of practices, operations, and circumstances. A cabinet of curiosities of the commonplace, the odd and sometimes the erroneous. A confession to strangeness, sometimes banal, sometimes intense.

This cabinet is a starting kit, a proposal for the reader to continue to think about and to collect similar examples. Which everyday weirdness is evoked by life with complex technologies and ungraspable techno-economic systems? How is your Mundania manifested? The curious, weird and incongruous is often hidden in the mundane. In that faint background hum. This is a suggestion to further unveil and capture the variations of Mundania. Because it might be in these openings, in the curiosities, that the potential lies. Instead of vanishing points, they can be vantage points to imagine differently, but to also question and scrutinize that which is unfair or outright wrong. The small stories of this cabinet might link to some of the discussions that have already been presented in the book. But here the stories are mostly intended to evoke openings in the shape of further question marks.

The chapter, or this small cabinet, will raise questions and instil more variation. How far can we stretch the Mundania-imaginary? The contrasting examples will point at the ambiguous nature of the variations of Mundania. Or are these examples fractures in the construct? Are these openings? To what? When is Mundania challenged by disruption, and when is mundanization still going on despite malfunctions, frictions, and discontent? What opens and what closes? For whom, how and when? There are variations in how Mundania can be experienced by different people. Uneven and often unfair circumstances. Mundanization is based on a tense interlink between enablement and submission. Between the calming and the creepy. Mundania is a realm of ambiguous amalgamations, experienced as comfortable and often pleasurable while upheld through sometimes wicked operations and the gradual acceptance of the uncanny.

Algorithmania

There is a soft sound in the kitchen. A process is going on while I am doing other things. The sound has been evoked daily for some time, but it has not always been there. It comes from a machine under the bench, next to the sink.

For many years we did not have any dishwasher at home. People who visited us often exclaimed: 'How can you cope with doing the dishes? It takes such a long time; it is so boring.' When, after some years, we bought a dishwasher, it began to take care of the procedure that had formerly been manual. It started to emit its gentle hum in the kitchen. Now dishwashing was based on programmes, held together by functional specifications. It became a clearly algorithmic process, it followed pre-specified rules and instructions in an automated procedure. Manual dishwashing had also been procedural, like so many habitual tasks. But with the machine, this procedural quality was augmented. If a button was pressed and a programme initiated, it should deliver a regular and expected result every time, if the maintenance of the machine had been conducted as described in the handbook and user guide.

The instalment of a dishwasher might be a banal shift for those who prefered and could afford this technological aid. For me, this shift turned out to be about more than changed routines. Through this transformation of dishwashing practices, I noticed a drift in my imaginaries and feelings about cleanliness and hygiene. It was not only that I could take the autopilot-stance, and let the machinery do my work. It was partly about filling the machine before running it, to optimize and to use energy in an effective way. But it was also about something more. I realized a shift in my imaginaries about what would be a proper wash. I wanted to run cutlery and kitchenware through the machine as part of quality assurance. The behaviour that emerged seemed to be built on the assumption that the run of things through the automatic dishwashing process was to some extent superior to what would happen if I were to wash up drinking glasses and plates manually. If I followed the algorithmic process things would become better. I stopped relying on my own competence and judgement. This was nothing dramatic, merely a slight nudge in the direction towards reliance on machinic processing. I realized that this happened to others. Some people told me, for example, that they run clothes through the washing machine, just in case. To fill the machine, but also to 'reboot' the clothes to a purer state. In this case, the washer/machine operator did not really examine whether the clothes were dirty and in need of washing before running them through the machine. Better machine processed than not. "Do we have something more to fill-up the machine?" When does this happen? This machinic cleansing and purgative ritual.

During the COVID-19 pandemic, air-filtering equipment got a similar machine-cleansing role. Air purifiers with HEPA-filters (high-efficiency particulate air [filter]), or similar technologies. To get fresh, processed and filtered air. New domestic appliances might become part of a new normal. Cleansing, and purification happening without us doing the cleaning, automatically, while we are busy doing something else. The processing of atmospheres is supplemented by devices monitoring the air quality. If the monitor says the air is good, then it is good. Doesn't the air smell fresher now? When do these products or services turn from add-ons of everyday life to necessities? When do machinic and algorithmic processes become imperative? Standardized and streamlined procedures, strengthened by beliefs in the unity of technologically generated systems, have a strong role in Swedish society. Better digitalized than not. They appear in everything from smart and connected homes to the implementation of automated decision making in the public sector.

How does the (over-)reliance on machinic processes occur? When are we supposed to run it all through the machine? Process it, give it machinic transformation. Automated, computed, regulated, standardized, with set and specified results. According to technical specifications. Preferences for machines are the prerequisites for much of digital transformation. But it goes beyond the digital, as a look at dishwasher procedures might show. Machines, not only computers and digital "solutions', seem to sometimes evoke the feeling that they deliver a quality that pertains merely to the fact that something has been run through the machine. Like the verification stamp that something gains when it has been run through an organizational process. When does this feeling of machinic substantiation emerge? How does *algorithmania* develop?

Swipe 'n' scroll

When I look at the Wi-fi router, I see flashing lights. Things work as they are supposed to do. The light indicates traffic. But what is traffic here? Vehicles, such as cars, are about traffic, and they have also become a problem to solve. Traffic congestions and global warming. Mobility technologies like the car have fostered societies in which traffic systems are a major feature. A central characteristic. Cities and other human habitats, as well as larger geographical areas, started to turn into trafficscapes after World War II. New phenomena occurred. Crossings and roadkills. Road sign interpretations and traffic cultures. Sound horn or give way. Speed limits and parking lots. Land as well as practices have been ordered according to principles of traffic for decades.

Marshall McLuhan brought up the shift to a car-centric society and proposed that there was also 'a growing uneasiness about the degree to which cars have become the real population of our cities, with a resulting loss of human scale, both in power and in distance' (McLuhan 1995: 218). McLuhan continued to propose that city planners begin to buy back cities for pedestrians from the big transportation interests. Shannon Mattern takes this as the point of departure to discuss the role of sidewalks (Mattern 2022). It is by shifting perspective from the main lane to the peripheries of a system that novel paradigms might be spotted. When scrutinizing sidewalks, now and before, we can see how negotiations between the public and the private have come about, as well as how different aspects of power and social dynamics can be noticed (Mattern 2022: 41). The sidewalk has been where acts of deference or domination have been negotiated, but also, 'sidewalks were sites of contestation and media for resistance' (Mattern 2022: 42). What will be the future technologies that have similar impact as cars, and can some of these even be spotted on sidewalks and maybe in domestic spaces? How will they be involved in negotiations between deference, domination and resistance?

If the car came to characterize urban spaces for decades, in domestic spaces, fridges, television and electric light have evoked new ecologies. What would be the equivalent to the car in a domestic space in the early 2020s, and what would be the sidewalk? Smartphones, computers and other screen-based devices have surely become prevalent to organize domestic life. Maybe the smartphone had become the car of homes, and computers the trucks? Steve Jobs, founder of Apple, actually likened devices with a touchscreen to cars, and personal computers (PCs) to trucks (Snell 2016).

With the personal or home computer and its graphical user interface based on the desktop metaphor, a certain ecology was introduced in households some decades ago. For years, office work became the underlying ordering principle for digital cultures. Folders and files. These are still prevalent, but with new devices and interfaces also came new orders. Files were moved into the background when other features popped up. Such as the Feed, as it appeared on people's screens. The never-ending flow or list of items in social media applications like Instagram, TikTok or X (Twitter). A new kind of traffic. Vertical scroll became a major movement of lives with smartphones. For people growing up with this, the teenage years were more characterized by swipe'n'scroll than rock'n'roll. A new geography of attention emerged. Through swipe'n'scroll, posts, videos, sounds and images constantly emerged from beneath, in a constant flow propelled by a repeated gentle thumb-swipe over the glass-covered touchscreen. New traffic. New ways to be moved.

According to the plan

'Black Week'. During the last years it has become part of the yearly consumption rhythm. A week of special offers, some weeks before Christmas. I look at a website for one of the Swedish telecom companies. Here are mobile phones on offer. iPhones, Samsung Galaxy, Sony Xperia, Xiaomi and Phone (1) by Nothing. It seems that I can subscribe to these? What does it mean?

There are different factors that set the pace in Mundania. Technological development is one such. Another is the rhythms of consumption, synchronized by business models. A key factor when using a smartphone in the early 2020s was to have it connected to the services of a telecom company, an operator providing network connection. Operators had different deals to offer phones + their own services. Often these deals or plans were based on subscriptions, on credit and monthly instalments. Not only for the service but also for the hardware. These deals, like several others based on future payments, encouraged people to buy (more) expensive things. Payment postponed and smeared out over the future. This was part of the further consolidation of ambient consumption.

What are the ends of what you can subscribe to? Some telecom companies, like Telenor in Sweden, not only sold hardware such as smartphones, but they also offered upgrade plans, not for software but for smartphones. When signing these deals, customers were able to exchange their present smartphone with a brand new one every 12 months (Telenor Change n.d.). This plan meant a new device every year, and instead of paying maybe around 10,000 Swedish crowns (around €900 in 2023) when receiving the new device, the payment was dissolved into future instalments that kept on gradually accumulating.

One purpose of these plans was to keep customers bound to services for as long as possible. Most people did not have plans that included a yearly upgrade of phones, but by simply signing up to any of the subscription agreements they also became part of the push towards increased ambient consumption. With the introduction of subscription also for hardware, the idea about purchasing as a single one-time transaction further dissolved. In this way, expensive and resource-demanding devices like smartphones, could become something that was regularly delivered, that just kept on coming as long as the customer was creditworthy.

According to this logic, there is a further strengthening of the vapourware phenomenon. Flawless products exist merely as vapourware. In a haze of hype and hope. In some future. Or in 12 months, when a better phone will arrive, automatically, according to the plan. Version 1.0 of something is followed by 1.1. It is expected to be better, 1.2 even better, and so on. The coming version will come with smaller or even larger improvements and fixes. Consumption can be driven by imaginaries about what is inadequate, or even about recurring failure.

Failure can be concrete and unanimous. Like a plane crash. It can, however, also be relative. Something that you have been doing or something that you possess or have had a longer relationship to begin to feel inferior, inadequate and wrong. Time to upgrade, time to change. According to Arjun Appadurai and Neta Alexander, failure should be related to the social dimensions of expectations. In this sense, failure can be understood as a judgement (2020: 2). They argue that there has emerged a logic within technology businesses, driven by the practices of Silicon Valley, that is based on recurrently unfulfilled promises (Appadurai and Alexander 2020: 20-21). When products are introduced, they are marketed as flawless products. Quite soon, when launched and adopted, they instead turn into short steps on the path towards a better future. A future supposed to be fulfilled by the upcoming versions of present things. A product becomes a process, a service, always in development, part of an upgrade cycle. Betterment forthcoming, hinted at as vapourware. Imaginaries about coming things being faster, better and frictionless influence how devices at hand are experienced. Imaginaries about forthcoming faster things make present things slow and the waiting feel longer (Appadurai and Alexander 2020; Willim 2003a).

Through the ephemerality of products, through this indirect planned obsolescence of devices, tensions occur. One such tension can be between things promoted as precious, as more-or-less timeless, that will also soon be outdated due to changed technological circumstances and the logics of business plans. A prime example is an extremely expensive phone or maybe smartwatch with much of its value in a luxurious physical appearance that is also equipped with electronic functionalities. The latter features quickly become dated and unusable, while the physical artefact can still be experienced as luxurious and enticing. It is like a Fabergé egg that, when opened, is hiding a set of outdated, defunct and cheap-looking electronic components and some software that no one can access anymore.

A further question is how more-or-less tantalizing devices relate to ideas about technologies as utilities? In early 2020s Sweden a smartphone was not something that came with extraordinary status; instead it had become a necessity, a utility needed to partake in much of what happened in society. A smartphone had become the nexus of logistical media, a key device in many respects. Like having a passport, bank account or credit card. The difference was that some could have a bit fancier and more advanced 'passports' than others. Fancy utilities that could be smoothly replaced yearly. All according to the plan. All while the flows and piles of e-waste increased.

Quick and contactless

It is October 2022, and I touch an icon on the smartphone screen. The app of a food vendor opens and presents its brisk design. Offerings from different restaurants pop up on the screen. Not just food and beverages, but also groceries, candy and electronics can be ordered. This is part of so-called

Q-Commerce, or Q-Comm, a business phenomenon that has grown during the last years. A description of it is offered on the webpage *swish.nu*.

> The journey from the consumer's interest being aroused to the product being paid for and delivered outside the door must go fast, and today we are talking in terms of thirty-minutes-fast! This is Quick Commerce, the new black in e-commerce, that has really taken off as a result of the Corona Pandemic. We explain what the new retail trend means, and explain why your choice of payment method may be as important as the choice of delivery method. (Swish website n.d.)

Who said this? Swish is the name of a payment service that was widely used in Sweden during the early 2020s. That is why the choice of payment method is emphasized in their description of Q-comm. 'Swish' is an onomatopoetic word that evokes that something is passing by quickly, and is similar to the English expression *swoosh*. They have also chosen the domain-name '.nu', that in Swedish spells the word *now*. The address of the site is basically *swoosh.now*.

Swish was started in 2012 as a cooperation between six of the largest banks in Sweden. The service is principally a system for mobile payment. To verify a payment, an identification method must be used. Such as Bank-ID, which was dominant in Sweden in the early 2020s. Bank-ID was also a collaboration between banks, and it was not just used for commercial transactions, but also for people to identify themselves when they log in to public services such as healthcare and the tax authorities. The financial world became very tightly interlinked with the public and with everyday life in Sweden during the early 2020s.

At this time Q-commerce was described as the third generation of e-commerce. The first generation was based on self-service systems in shops. The second generation on internet orders, with delivery times being around a couple of days. The third generation has a considerably quicker pace, with delivery times in less than an hour. Swish, swoosh, quick commerce. It is promoted as an inevitable progress, like an evolutionary path of acceleration.

Back to my phone. I distractedly click on some of the offerings in the food q-comm app. I realize that the design of the app has been upgraded since I last saw it. I come to think of the way that groceries are frequently reorganized in a physical shop. Moved to attract attention and to test how people react to the changes. Then I close the app. Without ordering anything. No Q-commerce this time. I put down my phone. Then after a while, the screen lights up again. Something wants my attention. A notice emerges. Jauntily it nudges me to open the app again. It wants to bring me back. Did I forget something, couldn't I decide what to order, did I miss some offerings and bargains? 'We are ready when you are + winking face emoji'.

If I then were to open the app, if I were to choose any of the offerings from any of the local shops or restaurants, a process would be initiated. Right after clicking the confirmation button, the chosen restaurant would get a ping with the order. They could confirm it and start preparing it. Now the order would emerge among the riders as a task. Among foremost bike-riders with oversized rectangular backpacks. The task would appear in their app, which is different than the one I have. The riders are strongly attached to their smartphones; these are the devices that coordinates, that help them communicate and orient themselves, meanwhile that they are tracked and monitored (Andersson 2021). Their performance is constantly measured, regulated and evaluated. When a rider accepts the task to deliver the order, a timer starts. Now it is a race against time to deliver as smoothly as possible. Smooth and quick, as in Q-commerce.

When ordering I could choose 'contactless delivery'. This became a normal during the pandemic. The possibility to order something; packages, food and items that are delivered on the doorstep or at the perimeter of addresses without any contact between the delivery person and the recipient. Contactless delivery was also widely used after the end of the pandemic. Streets in cities were busy with riders that swiftly moved between address and address, performing quick commerce and contactless delivery. Smooth services. Convenient on one end and demanding, taxing, arduous, on the other. Machines and humans as parts of a system. Like so many times before. Buttons are pressed, commands are conveyed, and stuff and people are put to work. The ends and beginnings of infrastructures, technologies and labour become more and more unclear. What fuels the acceleration? Speed requires labour and generally consumes energy. Who are all the stakeholders involved, what are their interests and rights? What are the implications of these kind of services that emerge in-between, that extend beyond and beneath? Which dimensions of Mundania emerge and get hidden in phenomena like Q-commerce and contactless delivery?

Browsing ratio

I have just registered for a one-month free subscription. Did I need another one? I registered my credentials, including a payment method. No money is withdrawn, but if I have not cancelled the subscription after a month, the rhythm of transactions will start ticking. One tick, one payment per month. This is a subscription service for magazines and newspapers. As long as I am part of the plan, I have access to 6,000 magazines and newspapers.

I download and open the app. I choose three categories of magazines that I am interested in. Then I get to the Discover-page. Colourful covers of magazines are presented in row after row. Titles from Australia, Germany, Sweden and so on. Presented for me according to some algorithmic logic. Orderings, lists and nudges. I swiftly find magazines that I feel are interesting.

I mark some as favourites. I start collecting. I let some suggestions pass by, also some recommendations for the most popular titles at the time. I browse among the presented magazines. Now and then I click on a cover. The magazine open and shows the first page of the content. I swipe to the left; I browse through the pages. I read some chunks of texts. Look at images. I continue browsing. Then I close the magazine, and I continue to instead browse the different categories in the app.

What is the *browsing ratio* here? How much time do I spend browsing, and how much time do I spend on reading 'content'. What is the content here, what is the medium, what is the primary experience? Has the app, the platform, taken priority over the items that are presented in it?

I recognize it all from my time spent with streamed music or video. The amount of time spent on discovering and searching for stuff. Browsing. Marking items as favourites, stuff to digest later. In the services for video, I have recognized that I collect titles in 'my list', but when I watch films or series they are often not from that list of favourites. There are no guilty pleasures in the list, no moving media fast-food. Maybe the list is an ideal mirror image of me as media consumer, a list full of nourishing and healthy stuff that I then do not digest that often. I guess that in the cupboards in the kitchen there are some products that are like the items in the video favourites list. Discovered and acquired long ago, collected for a healthier life, a future life. Then forgotten.

With the magazine subscription app, there is also another dimension. The idea about browsing among covers that can be opened for more browsing gives it all an extra dimension. Layers of browsing. What is the browsing ratio throughout different circumstances, formats and media in Mundania, and what is browsing?

To browse is to glance through, a kind of brief investigation or survey. It has some resemblances with scrolling. Minor and often repetitive movements that evoke items, information and sensations. Browsing came from printed media to screen navigation. When new ways to engage with media emerge, browsing might become a less obvious action. Navigation through certain interfaces such as links, menus, search boxes and various screen elements that can be touched has been central to life with digital technologies during the first decades of the 2000s. Gradually these practices merge with new ways of engagement, where conversations with systems become prevalent. New vanishing points emerge. More intimate and personal ways of engagement? What are the different moves and skills involved in these different practices? What is content and what is media, and how does it transform over time and between contexts?

Flat rate

Think about all the flashing lights of Mundania. The lamps and indicators on devices. They imply something. Sometimes they show the flow and traffic of something. Workings, status and levels. Sometimes they hint at underlying temporal conditions. Maybe at *Taximeter time*. This is a temporal condition that is spread right across clock-based societies, a specific condition pertaining to specific logistical media. All the periods of time that are based on a meter ticking in the background while a process is going on, a meter that counts the consumption of something. The meter that clicks in the staircase in the example in Chapter 2, 'Vanishing Points', is a clear example. When I press the power button, lights shine and the meter starts clicking. Electricity is consumed. Beneath a door in the basement, there are more meters. Owned by the electricity company. Meters locked by small white plastic seals to avoid manipulation. The meters show in numbers on a display, in real-time, how electricity is consumed in the households of the building. How many real and imagined meters are ticking audibly or inaudibly throughout the variations of Mundania?

The ticking of meters, taximeter time, can be somewhat stressful. Especially when prices are raised. This happened in 2022. An energy crisis had built up for some time and escalated during the Russian invasion of Ukraine. Several following events made the price of electricity rise considerably. The only ones who did not care were the consumers that had managed to get electricity deals with a set price some years ago. The set price, the flat rate, has been a way out of taximeter stress. If someone has a flat rate deal on electricity, mobile connection, streaming of music or whatever it might be, they do not have to worry about the price per item or about the rate and tempo of the ticking taximeter. If a person has the financial resources, a flat rate is something desirable. It can shift attention away from certain kinds of consumption and make it ambient and infrastructural. Consumers can be ready to pay more just to get the relief from thinking and worrying about the ticking of taximeters. It shifts behaviour in similar ways as open bars. Free drinks! Then people normally don't worry about the price and value of items. At the end of the night, when the guests have left, you will find the side-effects of the open-bar concept. There will be half-full glasses, beer bottles and cans all around. Some opened and still almost full. It is easy to leave a half-full glass somewhere and just get another one from the bar.

The flat rate seems like a way to reach escape velocity, a means to move above all the constraints of logistical media. This escape velocity is, however, a false vision. The gravitation of the planet sooner or later draws people back to very tangible and overt circumstances. What can be experienced as flat rate transcendence is merely related to higher infrastructural depth, where people do not notice all the layers below. Until something fails. The flat rate makes people oblivious to the infrastructural depth of certain kinds of consumption. What are the long-term effects of this unawareness?

Conditions for smoothness

I am down on my knees on the wooden floor in the living room, and I look through a small lens. Or I rather look at the screen of my smartphone that is connected to a lens, that is part of a small microscope that I have acquired. The lens is pointed towards the floor, and what appears is a brown and yellow irregular topography of peaks and valleys. The even floor has turned into an irregular scape. The regularity of the rectangular boards of the floor has become an intricate geography. This topography has always been there, but this is the first time I can experience it in this way. New images and imaginaries have emerged from the floor. From now on, I can imagine them every time I look at the ordered pattern of the rectangular floorboards.

When certain technologies are used, regularity and order can be experienced differently. Regularity and order are a question of scale. A cheap microscope can make an ordered surface look intricately messy. The ordinary becomes at once both fantastic and uncanny through this lens. What if we were always to perceive reality at this scale? How would we then experience and manage order?

In Mundania new technologies are gradually introduced and adopted. Everything from press buttons to data centres has transformed the world in intricate ways. New things have emerged, while others have become ignored. The geography of Mundania is strange, it is fantastic and it is uncanny. For various reasons that I have touched upon, people might keep on experiencing it as ordinary and taken-for-granted. Like a smooth and regular pattern of the floorboards that underpin everyday life. What are the conditions for smoothness?

Lastly

This short cabinet of curiosities is a proposal to see questions also as openings. It is a proposal to avoid the fear of the question mark. To instead stay curious, to learn more, to probe more and to keep asking new and better questions. There is no single simple answer to the intricacies and possible challenges of Mundania. No universal formula telling what to embrace or adopt. What should, however, be avoided are the lures that emerge during challenging times.

One lure often occurs in populist politics and in conspiracy theories and constructs. It comes through self-assured proponents that propagate simple answers to complex questions. The rhetoric of this lure normally includes some scapegoat to blame and simplistic solutions. Another lure comes with the promotion of all-encompassing product portfolio futures. When some stakeholder proposes that 'We and our products will take care of it all!', it might imply 'One system to rule them all!' A brandverse with the ambition to become a universe. It is seldom wise to put all your eggs in the same basket. Instead, it might be wise to keep on asking what should be connected, and to what? A third lure is about excessive conduct, and the belief that people must slavishly stay with established formats, methods, models, orders and protocols. The potential might instead be to approach what is beyond all templates, the irregularities among all the smoothness and optimization, the patches among the seamlessness.

Mundania as it has been put forth in this book is a proposal to imagine differently when it comes to everyday life with ever more complex technologies. A proposal to approach variations of Mundania as something that keeps on emerging, differently for different people. Sometimes the Mundania concept might be easy to agree with, sometimes not. Maybe it is time to transform it or to try to see how to go further along with it? It is up to the reader to consider how useful Mundania is. When does it resonate with the world you live in? When does it just feel bothering and awkward? Here will be no clear answers. Instead, more question marks, and a plea for curiosity and also playfulness. Even in dire times. It means that some imagination is required, some assembly is required.

References

Ahmed, S. (2006) *Queer Phenomenology: Orientations, Objects, Others*. Durham, NC: Duke University Press.

Andersen, C.U. and Pold, S.B. (2021) The Metainterface Spectacle. In: *Electronic Book Review*. 6 November. http://electronicbookreview.com/essay/the-metainterface-spectacle/ (accessed November 2022).

Anderson, B. (1983) *Imagined Communities: Reflections on the Origin and Spread of Nationalism*. London: Verso.

Andersson, M. (2021) The Food Courier and His/Her Mobile Phone. In: A. Hill, M. Hartmann and M. Andersson (Eds) *The Routledge Handbook of Mobile Socialities* (pp 195–208). Oxon: Routledge.

Appadurai, A. (1997) *Modernity at Large*. New Delhi: Oxford University Press.

Appadurai, A. and Alexander, N. (2020) *Failure*. Cambridge: Polity Press.

Apple Support (2022) About Face ID advanced technology. https://support.apple.com/en-us/HT208108 (accessed March 2023).

Apple Support (2022) What's in The Menu Bar? (macOS Monterey 12). https://support.apple.com/en-gb/guide/mac-help/mchlp1446/mac (accessed August 2022).

Atkinson, P. (2013) *Delete: A Design History of Computer Vapourware*. London: Bloomsbury Academic.

Barlow, J.P. (1996) *A Declaration of the Independence of Cyberspace*. Electronic Frontier Foundation. Available at: www.eff.org/cyberspace-independence (accessed August 2023).

Bateson, G. (1972/1987) *Steps to an Ecology of Mind. Collected Essays in Anthropology, Psychiatry, Evolution, and Epistemology*. Northvale, NJ: Jason Aronson Inc.

Beyes, T. and Pias, C. (2019) The Media Arcane. *Grey Room* 75(Spring): 84–105.

Berg, M. (2022) Digital Technography: A Methodology for Interrogating Emerging Digital Technologies and Their Futures. *Qualitative Inquiry* 28(7): 827–836.

Berker, T., Hartmann, M. and Punie, Y. (206) *Domestication of Media and Technology*. London: Open University Press.

Bille, M. (2019) *Homely Atmosphere and Lighting Technologies in Denmark: Living with Light*. London: Bloomsbury Academic.

Bille, M., Bjerregaard, P. and Sørensen, T.-F. (2015) Staging Atmospheres: Materiality, Culture, and the Texture of the in-Between. *Emotion, Space and Society* 15: 31–38.

Björnberg, A. (2009) Learning to Listen to Perfect Sound: Hi-fi Culture and Changes in Modes of Listening 1950–1980. In: D. B. Scott (Ed) *The Ashgate Research Companion to Popular Musicology* (pp 105–129). London: Ashgate.

Blue, S. (2019) Institutional Rhythms: Combining Practice Theory and Rhythmanalyzis to Conceptualize Processes of Institutionalization. *Time & Society* 28(3): 922–950.

Boellstorff, T. (2015) *Coming of Age in Second Life: An Anthropologist Explores the Virtually Human*. Princeton: Princeton University Press.

Bolter, J.D. (2006) The Desire for Transparency in an Era of Hybridity. *Leonardo* 39(2): 109–111.

Bolter, J.D. and Grusin, R. (2000) *Remediation: Understanding New Media*. Cambridge, MA: MIT Press.

Bowker, G.C. and Star, S.L. (1999) *Sorting Things Out: Classification and Its Consequences*. Cambridge, MA: MIT Press.

Bowker, G.C., Baker, K., Millerand, F. and Ribes, D. (2010) Toward Information Infrastructure Studies: Ways of Knowing in a Networked Environment. In: J. Hunsinger, L. Klastrup and M. Allen (Eds) *International Handbook of Internet Research* (pp 97–117). Dordrecht: Springer Netherlands.

Boyd, C.P. and Edwardes, C. (Eds) (2019) *Non-Representational Theory and the Creative Arts*. Singapore: Palgrave Macmillan.

Bowker, G.C. (2020) Commentary: Reading the Endless Archive. In: R. Coover (Ed) *The Digital Imaginary: Literature and Cinema of the Database*. Bloomsbury. http://www.bloomsburycollections.com/book/the-digital-imaginary-literature-and-cinema-of-the-database, pp 95–100 (accessed 10 December 2021).

Böhme, G. (2014) The Theory of Atmospheres and its Applications (translated by A.-Chr. Engels-Schwarzpaul), *Interstices* 15: 92–99.

Böhme, G. (2017) *The Aesthetics of Atmospheres*. Ed. J.-P. Thibaud, London: Routledge.

Bridle, J. (2018) *New Dark Age. Technology and The End of the Future*. London: Verso Books.

Bringéus, N.-A. (1976/1986) *Människan som kulturvarelse*. Malmö: Liber förlag.

Britannica (2023) Moore's law. https://www.britannica.com/technology/Moores-law (accessed May 2023).

Broussard, M. (2023) *More Than a Glitch: Confronting Race, Gender, and Ability Bias in Tech*. Cambridge, MA: MIT Press.

Brunnström, L. (2018) *Swedish Design: A History*. London: Bloomsbury Publishing.

Burrell, J. (2016) How the Machine 'Thinks': Understanding Opacity in Machine Learning Algorithms. *Big Data & Society* 3(1): 1–12.

Canales, J. and Herscher, A. (2005) Criminal Skins: Tattoos and Modern Architecture in the Work of Adolf Loos. *Architectural History* 48: 235–256.

Case, J.A. (2013) Logistical Media: Fragments from Radar's Prehistory. *Canadian Journal of Communication* 38(3): 379–395.

Champion, M.S. (2019) The History of Temporalities: An Introduction. *Past & Present* 243(1): 247–254.

Chan, J., Pun, N. and Selden, M. (2013) The Politics of Global Production: Apple, Foxconn and China's New Working Class. *New Technology, Work and Employment* 28(2): 100–115.

Chun, W.H.K. (2016) *Updating to Remain the Same: Habitual New Media.* Cambridge, MA: MIT Press.

Coupland, D. (1995) *Microserfs.* New York: Harper Collins.

Crapanzano, V. (2004) *Imaginative Horizons: An Essay in Literary-philosophical Anthropology.* Chicago, IL: University of Chicago Press.

Crawford, K. and Joler V. (2018) *Anatomy of an AI System: The Amazon Echo as an Anatomical Map of Human Labor, Data and Planetary Resources.* https://anatomyof.ai (accessed 29 April 2019).

Czarniawska, B. (2019) Virtual Red Tape, Or Digital v. Paper Bureaucracy. In: B. Czarniawska and O. Löfgren (Ed) *Overwhelmed by Overflows? How People and Organizations Create and Manage Excess* (pp 170–190). Lund: Lund University Press.

Dahlgren, P. and Hill, A. (2022) *Media Engagement.* London: Routledge.

De Maistre, X. (1871) *A Journey Round My Room.* A Gutenberg E-book, Release date: 29 June 2020 [EBook #62519].

Dharampal, M. (2021) *A Theory of Digital Hygiene.* Institute of Network Cultures. https://networkcultures.org/blog/2021/05/12/digital-hygiene/ (accessed February 2022).

Douglas-Jones, R. (2020) On The Heaviness of Clouds: IoT Computing Through Fog, Mist and Dew. In: D. Jørgensen and F. A. Jørgensen (Eds) *Silver Linings. Clouds in Art and Science* (pp 232–245). Trondheim: Museumsforlaget.

Dourish, P. (2016) Rematerializing the Platform: Emulation and the Digital–Material. In: S. Pink, E. Ardèvol and D. Lanzeni (Eds) *Digital Materialities. Design and Anthropology* (pp 29–44). London: Bloomsbury.

Dunne A. and Raby F. (2013) *Speculative Everything: Design, Fiction, and Social Dreaming.* Cambridge, MA: MIT Press.

Eco, U. (1980/2004) *The Name of The Rose.* New York: Vintage.

Edwards, P.N. (2019) Infrastructuration: On Habits, Norms and Routines as Elements of Infrastructure. In: M. Kornberger M, G.C. Bowker, J. Elyachar, A. Mennicken, P. Miller, J.R. Nucho and N. Pollock (Eds) *Thinking Infrastructure* (pp 355–66). Bingley: Emerald Publishing Limited.

Egard, H. and Hansson, K. (2021) The Digital Society Comes Sneaking In. An Emerging Field and Its Disabling Barriers. *Disability & Society*, 23 August: 1–15.

Egard, H., Hansson, K. and Wästerfors, D. (Eds) (2022) *Accessibility Denied: Understanding Inaccessibility and Everyday Resistance to Inclusion for Persons with Disabilities*. London: Routledge.

Eggers, D. (2013) *The Circle*. London: Penguin Books.

Ehn, B. and Löfgren, O. (2010) *The Secret World of Doing Nothing*. Berkeley, CA: University of California Press.

Ehn, B., Löfgren, O. and Wilk, R. (2015) *Exploring Everyday Life: Strategies for Ethnography and Cultural Analysis*. Lanham, MD: Rowman & Littlefield.

Eriksen, T.H. (2016) *Overheating: An Anthropology of Accelerated Change*. London: Pluto Press.

Eriksen, T.H. (2021) Standing Still at Full Speed: Sports in an Overheated World. In: *Frontiers in Sports and Active Living* 3, May, 1–9.

Ericson, S. and Riegert, K. (Eds) (2010) *Media Houses. Architecture, Media, and the Production of Centrality*. New York: Peter Lang.

Ericson, S., Riegert, K. and Åker, P. (2010) Introduction. In: S. Ericson and K. Riegert (Eds) *Media Houses. Architecture, Media, and the Production of Centrality* (pp 1–18). New York: Peter Lang.

Flyte Co-pilot (n.d.) https://flytestore.com/products/co-pilot (accessed October 2023).

Flyte Instruction Manual (n.d.) https://cdn.shopify.com/s/files/1/1000/7716/files/Flyte_Manual_Warranty_9bdcf73b-a210-4114-a1e0-0611d47e6675.pdf?4308633266745485960 (accessed December 2022).

Flyte (n.d.) Story Walnut https://flytestore.com/products/58d3ce96ab51ae1200104182 (accessed May 2023).

Flyte website (n.d.) https://flytestore.com (accessed October 2023).

Folkbokföringslag (1991: 481) https://www.riksdagen.se/sv/dokument-och-lagar/dokument/svensk-forfattningssamling/folkbokforingslag-1991481_sfs-1991-481/ (accessed October 2023).

Fors, V., Pink, S., Berg, M. and O'Dell, T. (Eds) (2019) *Imagining Personal Data: Experiences of Self-Tracking* (1st edition). New York: Bloomsbury Academic.

Foster + Partners, Apple Park (2018) https://www.fosterandpartners.com/projects/apple-park (accessed May 2023).

Foster + Partners, Apple Visitor Center (2017) https://www.fosterandpartners.com/projects/apple-park-visitor-center (accessed May 2023).

Friedner, M. and Helmreich S. (2012) Sound Studies Meets Deaf Studies. *The Senses and Society* 7(1): 72–86.

Galloway, A.R. (2012) *The Interface Effect*. Cambridge: Polity Press.

Gams, M., Yu-Hua Gu, I., Härmä, A., Muñoz, A. and Tam, V. (2019) Artificial Intelligence and Ambient Intelligence. *Journal of Ambient Intelligence and Smart Environments* 11(1): 71–86.

Gane, N. (2021) Nudge Economics as Libertarian Paternalism. *Theory, Culture & Society* 38(6): 119–142.

Gartner Hype Cycle (n.d.) https://www.gartner.com/en/research/methodologies/gartner-hype-cycle (accessed August 2023).

Gaver, W., Boucher, A., Pennington, S. and Walker, B. (2004). Cultural Probes and the Value of Uncertainty. *Interactions – Funology* 11(5): 53–56.

Gernsback, H. (1925) The Isolator. *Science and Invention* 13(3): 214 and 281.

Gibson, W. (1982) Burning Chrome. *Omni* July: 72–107.

Gibson, W. (1984) *Neuromancer*. New York: Ace Books.

Google Data Centers (n.d.) https://www.google.com/about/datacenters/ (accessed May 2023).

Graham, S. and Marvin, S. (2001) *Splintering Urbanism: Networked Infrastructures, Technological Mobilities and the Urban Condition*. New York: Routledge.

Greenfield, A. (2006) *Everyware: The Dawning Age of Ubiquitous Computing*. San Francisco, CA: New Riders Publishing.

Greenfield, A. (2017) *Radical Technologies: The Design of Everyday Life*. London: Verso Books.

Grimeton history (n.d.) https://grimeton.org/en/history/ (accessed August 2023).

Gunn, W., Otto, T. and Smith, R.C. (2013) *Design Anthropology: Theory and Practice* (1st edn). Oxon: Routledge.

HaDEA (2023) *European Health and Digital Executive Agency (HaDEA) Digital Product Passport*, 2 May 2023. https://hadea.ec.europa.eu/calls-proposals/digital-product-passport_en#description (accessed May 2023).

Hadler, F. (2018) Beyond UX. *Interface Critique Journal* 1: 2–9. DOI: 10.11588/ic.2018.0.45695.

Hagood, M. (2019) *Hush: Media and Sonic Self-Control*. Durham, NC: Duke University Press.

Haider, J. and Sundin, O. (2019) *Invisible Search and Online Search Engines: The Ubiquity of Search in Everyday Life*. London: Routledge.

Harris, M. and Rapport, N. (2015) *Reflections on Imagination: Human Capacity and Ethnographic Method*. Farnham: Ashgate Publishing.

Heckman, D. (2008) *A Small World: Smart Houses and the Dream of the Perfect Day*. Durham, NC: Duke University Press.

Highmore, B. (2001) *Everyday Life and Cultural Theory*. London: Routledge.

Hildebrand, J.M. (2020) Drone Mobilities and Auto-Technography. In: M. Büscher, M. Freudendal-Pedersen, S. Kesselring and N. Grauslund Kristensen (Eds) *Handbook of Research Methods and Applications for Mobilities* (pp 92–101). Cheltenham: Edward Elgar Publishing.

Hu, T.-H. (2015) *A Prehistory of the Cloud*. Cambridge, MA: MIT Press.

Husz, O. (2022) *The Identity Economy*. Göteborg: Makadam förlag (Part of *After Digitalisation* the 2022 Yearbook for Riksbankens Jubileumsfond).

Ingold, T. (2011) *Being Alive: Essays on Movement, Knowledge and Description.* Oxon: Routledge.

Ingold, T. (2014) 'That's Enough about Ethnography!'. *HAU: Journal of Ethnographic Theory* 4(1): 383.

Ingold, T. (2018) From Science to Art and Back Again: The Pendulum of an Anthropologist. *Interdisciplinary Science Reviews* 43(3–4), 2 October 2018: 213–227.

Internetmuseum, Carl Bildt och Bill Clinton skriver historia med sitt mejlande (2014) 4 December. https://www.internetmuseum.se/tidslinjen/bildt-och-clinton-mejlar/) (accessed May 2023).

Jackson, S.J. (2014) Rethinking Repair. In: T. Gillespie, P.J. Boczkowski and K.A. Foot (Eds) *Media Technologies: Essays on Communication, Materiality, and Society* (pp 221–239). Cambridge, MA: MIT Press.

Jakobsson, P. and Stiernstedt, F. (2010) Googleplex and Informational Culture. In S. Ericson and K. Riegert (Eds) *Media Houses. Architecture, Media, and the Production of Centrality* (pp 113–135). New York: Peter Lang.

Jalas, M. and Rinkinen, J. (2016) Stacking Wood and Staying Warm: Time, Temporality and Housework around Domestic Heating Systems. *Journal of Consumer Culture* 16(1): 43–60.

Jasanoff, S. (2015) Future Imperfect: Science, Technology, and the Imaginations of Modernity. In: S. Jasanoff and S.-H. Kim (Eds) *Dreamscapes of Modernity: Sociotechnical Imaginaries and the Fabrication of Power* (pp 1–33). Chicago, IL: The University of Chicago Press .

Jasanoff, S. (2021) The Dangerous Appeal of Tech. *MIT Technology Review* 124(4): 16–17.

Jaumotte, F., Li, L., Medici, A., Oikonomou, M., Pizzinelli, C., Shibata, I., et al (2023) Digitalization during the COVID-19 Crisis: Implications for Productivity and Labor Markets in Advanced Economies. International Monetary Fund. Staff Discussion Note SDN2023/003. International Monetary Fund, Washington DC.

Jönsson, L.-E. (2017) Byggnaderna, sakerna, arkivakterna. In: L.-E. Jönsson and F. Nilsson (Eds) *Kulturhistoria: En etnologisk metodbok* (pp 73–87). Lund: Department of Arts and Cultural Sciences, Lund University.

Karlsson, J. (2022) Mystiskt ljusfenomen misstogs för meteorregn – troligtvis en raketuppskjutning. *SVT Nyheter.* 13 August. https://www.svt.se/nyheter/inrikes/mystiskt-ljusfenomen-misstogs-for-meteorregn (accessed May 2023).

Kelly, K. (1999) *New Rules for the New Economy: Ten Radical Strategies for a Connected World.* New York: Penguin.

Kien, G. (2008) Technography = Technology + Ethnography. *Qualitative Inquiry* 14(7): 1101–9.

Kitchin, R. and Dodge, M. (2019) The (In)Security of Smart Cities: Vulnerabilities, Risks, Mitigation, and Prevention. *Journal of Urban Technology* 26(2): 47–65.

Kitnick, A. (2011) This is Marshall McLuhan. *Rhizome* 2011-10-04. https://rhizome.org/editorial/2011/oct/04/marshall-mcluhan/ (accessed August 2023).

Klingmann, A. (2007) *Brandscapes. Architecture in the Experience Economy.* Cambridge, MA: MIT Press.

Klinkmann, S.-E. (2005) Synch/Unsynch. *Ethnologia Europaea. Journal of European Ethnology* 35(1): 81–87.

Krapp, P. (2011) *Noise Channels: Glitch and Error in Digital Culture.* Minneapolis, MN: University of Minnesota Press.

Larsson, S. (2013) Metaphors, Law and Digital Phenomena: The Swedish Pirate Bay Court Case. *International Journal of Law and Information Technology* 21(4): 354–379.

Larsson, S. and Andersson Schwartz, J. (Eds) (2018). *Developing Platform Economies: A European Policy Landscape.* Brussels and Stockholm: European Liberal Forum asbl.

Latour, B. (1987) *Science in Action. How to Follow Scientists and Engineers through Society.* Cambridge, MA: Harvard University Press.

Latour, B. (2004) Why Has Critique Run out of Steam? From Matters of Fact to Matters of Concern. *Critical Inquiry* 30(2): 225–248.

Law, J. and Ruppert, E. (2013) The Social Life of Methods: Devices. *Journal of Cultural Economy* 6(3): 229–240.

Laws of UX, Tesler's Law (n.d.) https://lawsofux.com/teslers-law/ (accessed May 2022).

Lehmuskallio, A. and Meyer, R. (2022) Experimental Indices: Situational Assemblages of Facial Recognition. *The Journal of Media Art Study and Theory* 3(1): 85–112.

Lessig, L. (1999) *Code and Other Laws of Cyberspace.* New York: Basic Books.

Ling, R. (2012) *Taken for Grantedness. The Embedding of Mobile Communication into Society.* Cambridge, MA: MIT Press.

Löfgren, O. (2005) Catwalking and Coolhunting: The Production of Newness. In: O. Löfgren and R. Willim (Eds), *Magic, Culture and the New Economy.* Oxford: Berg.

Löfgren, O. (2014) At the Ethnologists' Ball: Changing an Academic Habitus. *Ethnologia Europaea: Journal of European Ethnology* 44(2): 116–122.

Löfgren, O. (2015) The Black Box of Everyday Life. Entanglements of Stuff, Affects, and Activities. *Cultural Analysis* 13, 77–98.

Löfgren, O. (2015b) The Scholar as Squirrel: Everyday Collecting in Academia. *Hamburger Journal Für Kulturanthropologie (HJK)* 3: 17–33.

Löfgren, O. (2017) Mess: On Domestic Overflows. *Consumption Markets & Culture* 20(1): 1–6.

Löfgren, O. (2021) Media and Mood Work: Routines, Daydreams and Micro-Moves. In: A. Hill, A.M. Hartmann and M. Andersson (Eds) *The Routledge Handbook of Mobile Socialities* (pp 38–52). Oxon: Routledge.

Löfgren, O. and Willim, R. (2005) *Magic, Culture and The New Economy*. Oxford: Berg.

Loos, A. (1908) Ornament and Crime, in German as Ornament und Verbrechen. https://de.wikisource.org/wiki/Ornament_und_Verbrechen (accessed August 2023).

Lorimer, H. (2005) Cultural Geography: The Busyness of Being 'More-than- Representational'. *Progress in Human Geography* 29(1): 83–94.

Lovink, G. (2019) *Sad by Design: On Platform Nihilism*. London: Pluto Press.

Lury, C. and Wakeford, N. (2014) Introduction: A Perpetual Inventory. In C. Lury and N. Wakeford (Eds) *Inventive Methods: The Happening of The Social* (pp 1–24). London: Routledge.

Mager, A. and Katzenbach, C. (2021) Future Imaginaries in the Making and Governing of Digital Technology: Multiple, Contested, Commodified, *New Media & Society* 23(2): 223–36.

Mahesha, R. (2018) How Cloud, Fog, and Mist Computing Can Work Together. *IBM Developer*. https://developer.ibm.com/articles/how-cloud-fog-and-mist-computing-can-work-together/ (accessed May 2022).

Manovich, L. (2001) *The Language of New Media*. Cambridge, MA: MIT Press.

Manovich, L. (2013) *Software Takes Command*. New York: Bloomsbury Academic.

Markham, A. (2013) Remix Culture, Remix Methods: Reframing Qualitative Inquiry for Social Media Contexts. In: N.K. Denzin and M.D. Giardina (Eds) *Global Dimensions of Qualitative Inquiry* (pp 63–81). Walnut Creek, CA: Left Coast Press.

Markham, A. (2020) The Limits of the Imaginary: Challenges to Intervening in Future Speculations of Memory, Data, and Algorithms. *New Media & Society* 23(2): 382–405.

Martínez, F. and Laviolette, P. (2019) *Repair, Brokenness, Breakthrough. Ethnographic Responses*. Oxford: Berghahn.

Mattern, S. (2017) *Code and Clay, Data and Dirt: Five Thousand Years of Urban Media*. Minneapolis, MN: University of Minnesota Press.

Mattern, S. (2022) Sidewalks of Concrete and Code. In: S. Sharma and R. Singh (Eds) *Re-Understanding Media: Feminist Extensions of Marshall McLuhan* (pp 36–50). New York: Duke University Press.

McLuhan, M. (1964/1994) *Understanding Media: The Extensions of Man*. Cambridge, MA: MIT Press.

McLuhan, M. and Carson, D. (2003) *The Book of Probes*. Berkeley, CA: Gingko Press GmbH.

McQuire, S. (2003) From Glass Architecture to Big Brother: Scenes from a Cultural History of Transparency. *Cultural Studies Review* 9(1): 103–123.

Merleau-Ponty, M. (1945/2012) *Phenomenology of Perception*. London: Routledge.

Michael, M. (2016) Notes toward a Speculative Methodology of Everyday Life. *Qualitative Research* 16(6): 646–660.

Microsoft 365 Support (n.d.) Turn the snap to grid and snap to object options on or off in Excel. https://support.microsoft.com/en-us/office/turn-the-snap-to-grid-and-snap-to-object-options-on-or-off-in-excel-817da2d4-40cc-4cff-b0de-1359ecbe95e3 (accessed May 2023).

Microsoft News (n.d.) Launch of Windows 95. https://news.microsoft.com/announcement/launch-of-windows-95/ (accessed December 2022).

Mol, A. (2002) *The Body Multiple: Ontology in Medical Practice*. Durham, NC: Duke University Press.

Moores, S. (2021) Investigating "Communities of Co-Movers": Motricity, Spatiality and Sequentiality in Social Life. In: A. Hill, M. Hartmann and M. Andersson (Eds) *The Routledge Handbook of Mobile Socialities* (pp 53–68). Oxon: Routledge.

Morozov, E. (2013) *To Save Everything, Click Here: Technology, Solutionism and the Urge to Fix Problems That Don't Exist*. London: Penguin Books.

mui Board (n.d.) https://muilab.com/en/products_and_services/muiboard/ (accessed August 2023).

Mullaney, T.S., Peters, B., Hicks, M. and Philip, K. (Eds) (2021) *Your Computer Is on Fire*. Cambridge, MA: MIT Press.

Mulvin, D. (2021) *Proxies: The Cultural Work of Standing In*. Cambridge, MA: MIT Press.

Nilsson, G. (2013) Balls Enough: Manliness and Legitimated Violence in Hell's Kitchen, *Gender, Work & Organization* 20(6): 647–663.

Noble, S.U. (2018) *Algorithms of Oppression: How Search Engines Reinforce Racism*. New York: NYU Press.

Norman, D.A. (1998) *The Design of Everyday Things*. Cambridge, MA: MIT Press.

Norman, D.A. (2010) *Living with Complexity*. Cambridge, MA: MIT Press.

Nye, D.E. (1996) *American Technological Sublime*. Cambridge, MA: MIT Press.

Nye, D.E. (2018) *American Illuminations. Urban Lighting, 1800–1920*. Cambridge MA: MIT Press.

O'Dell, T. and Willim, R. (2011a) Irregular Ethnographies: An Introduction. *Ethnologia Ethnology* 41(1): 26–39.

O'Dell, T. and Willim, R. (2013) Transcription and the Senses: Cultural Analysis When It Entails More Than Words. *Senses and Society* 8(3): 314–334.

O'Dell, T. and Willim, R. (2017) Entanglements: Issues in Applied Research and Theoretical Scholarship. In: S. Pink, V. Fors, V. and T. O'Dell (Eds) *Theoretical Scholarship and Applied Practice* (pp 206–224). Oxford: Berghahn.

Orwell, G. (1949) *Nineteen Eighty-Four*. London: Secker & Warburg.

Oxfam (2022) Global Index Shows Sweden Worst in the Nordic Countries at Fighting Inequality. 11 October 2022. https://oxfam.se/en/nyheter/globalt-index-visar-sverige-samst-i-norden-pa-att-bekampa-ojamlikhet/ (accessed May 2023).

Paasonen, S. (2021) *Dependent, Distracted, Bored. Affective Formations in Networked Media*. Cambridge, MA: MIT Press.

Parikka, J. (2015) *A Geology of Media*. Minneapolis, MN: The University of Minnesota Press.

Parikka, J. (2017) The Sensed Smog: Smart Ubiquitous Cities and the Sensorial Body. *The Fibreculture Journal* (29) FCJ–219. doi: 10.15307/fcj.29.219.2017

Parisi, D. (2018) *Archaeologies of Touch: Interfacing with Haptics from Electricity to Computing*. Minneapolis, MN: Minnesota University Press.

Parker, M. (2020) The People's Cloud: A Sonospheric Investigation. In: D. Jørgensen and F.A. Jørgensen (Eds) *Silver Linings: Clouds in Art and Science* (pp 226–231). Trondheim: Museumsforlaget.

Parks, L. (2009) Around the Antenna Tree: The Politics of Infrastructural Visibility. *Flow Journal*. 6 March. https://www.flowjournal.org/2009/03/around-the-antenna-tree-the-politics-of-infrastructural-visibilitylisa-parks-uc-santa-barbara/ (accessed May 2023).

Pasquale, F. (2015) *The Black Box Society: The Secret Algorithms That Control Money and Information*. Cambridge, MA: Harvard University Press.

Peters, J.D. (2015) *The Marvelous Clouds. Toward a Philosophy of Elemental Media*. Chicago, IL: The University of Chicago Press.

Petersén, M. (2019) *The Swedish Microchipping Phenomenon*. Bingley: Emerald Publishing Limited.

Petrick, E.R. (2020) Building the Black Box: Cyberneticians and Complex Systems. *Science, Technology, & Human Values* 45(4): 575–595.

Pink, S. (2020) #stayathome: Being in an Uncertain Place. In: *Future Matters*, Melbourne: ETLab, Monash University.

Pink, S. (2022) *Emerging Technologies / Life at the Edge of the Future*. London: Routledge.

Pink, S., Ruckenstein, M., Willim, R. and Duque, M. (2018) Broken Data: Conceptualising Data in an Emerging World. *Big Data & Society* 5(1): 1–13.

Pink, S., Fors, V., Lanzeni, D., Duque, M., Sumartojo, S. and Strengers, Y. (2022) *Design Ethnography: Research, Responsibilities, and Futures* (1st edn). London: Routledge.

Plotnick, R. (2018) *Power Button: A History of Pleasure, Panic, and the Politics of Pushing*. Cambridge, MA: MIT Press.

Pogue, D. (2013) Apple Shouldn't Make Software Look Like Real Objects. *Scientific American*. 1 February. https://www.scientificamerican.com/article/apple-shouldnt-make-software-look-like-real-objects/ (accessed November 2022).

Purify app (n.d.) https://www.purify-app.com (accessed May 2023).

Rahm, L. and Kaun, A. (2022) Imagining Mundane Automation: Historical Trajectories of Meaning Making Around Technological Change. In: S. Pink, M. Berg, D. Lupton and M. Ruckenstein (Eds) *Everyday Automation: Experiencing and Anticipating Emerging Technologies* (pp 23–43). London: Routledge.

Rasmussen, E.D. (2020) Clouds of Unknowing in Don DeLillo's Underworld. In: D. Jørgensen and F.A. Jørgensen (Eds) *Silver Linings. Clouds in Art and Science* (pp 200–209). Trondheim: Museumsforlaget.

Reddigari, M. (2022) 8 Genius Ways to Hide Every Wire in Your Home. In: *Bob Vila: Tried, True, Trustworthy Home Advice*. 26 October. https://www.bobvila.com/articles/how-to-hide-tv-wires/ (accessed May 2023).

Rees, T. (2018) *After Ethnos*. Durham, NC: Duke University Press.

Regeringens skrivelse 2017/18:47 (2017). Hur Sverige blir bäst i världen på att använda digitalizeringens möjligheter – en skrivelse om politikens inriktning.

Reid, G. (1999) Of Responses & Resonance. *Sound on Sound*. October. https://www.soundonsound.com/techniques/responses-resonance (accessed August 2023).

Reyes, I. (2010) To Know Beyond Listening: Monitoring Digital Music. *The Senses and Society* 5(3): 322–338.

Robertson, C. (2009) A Documentary Regime of Verification: The Emergence of the US Passport and the Archival Problematization of Identity. *Cultural Studies* 23(3): 329–354.

Rogers, E. (1962) *Diffusion of Innovation*. New York: Free Press of Glencoe.

Rose, S. (2013) Why Apple Ditched Its Skeuomorphic Design for IOS7. *The Guardian*. 12 June. https://www.theguardian.com/technology/shortcuts/2013/jun/12/skeuomorphism-apple-ditched-ios7 (accessed November 2022).

Rosner, V. (2020) *Machines for Living: Modernism and Domestic Life*. Oxford: Oxford University Press.

Rossiter, N. (2021) Logistical Media Theory, the Politics of Time, and the Geopolitics of Automation. In: M. Hockenberry, N. Starosielski and S. Zieger (Eds) *Assembly Codes: The Logistics of Media* (pp 132–50). Durham, NC: Duke University Press.

Rotman, D. (2020) We're not prepared for the end of Moore's law. *MIT Technology Review*, 24 February 2020. https://www.technologyreview.com/2020/02/24/905789/were-not-prepared-for-the-end-of-moores-law/ (accessed August 2023).

Sadowski, J. (2020a) The Internet of Landlords: Digital Platforms and New Mechanisms of Rentier Capitalism. *Antipode* 52(2): 562–580.

Sadowski, J. (2020b) *Too Smart: How Digital Capitalism is Extracting Data, Controlling Our Lives, and Taking Over the World*. Cambridge, MA: MIT Press.

Scheerbart, P. (2014) Glass Architecture, 1914. In: J. McElheny and C. Burgin (Eds) *Glass! Love!! Perpetual Motion!!!: A Paul Scheerbart Reader* (pp 20–91). Chicago, IL: University of Chicago Press.

Schumpeter, J.A. (1994) *Capitalism, Socialism and Democracy*. London: Routledge.

Schwab, K. (2019) *The Global Competitiveness Report 2019*. Geneva: World Economic Forum.

Seaver, N. (2022) *Computing Taste: Algorithms and the Makers of Music Recommendation*. Chicago, IL: University of Chicago Press.

Sharma, S. and Singh, R. (Eds) (2022) *Re-Understanding Media. Feminist Extensions of Marshall McLuhan*. Durham, NC: Duke University Press.

Shove, E. (2003) *Comfort, Cleanliness and Convenience: The Social Organization of Normality*. Oxford: Berg Publishers.

Silverstone, R., Hirsch, E. and Morley, D. (1992) Information and Communication Technologies and the moral economy of the household. In: R. Silverstone and E. Hirsch (Eds) *Consuming Technologies: Media and Information in Domestic Spaces* (pp 115–131). London: Routledge.

Simmel, G. (1906) The Sociology of Secrecy and of Secret Societies. *American Journal of Sociology* 11(4): 441–498.

Sneath, D., Holbraad, M. and Pedersen, M.A. (2009) Technologies of the Imagination: An Introduction. *Ethnos* 74(1): 5–30.

Snell, J. (2016) Cars, Trucks, iPads, and Laptops: When the Touchscreen Generation Would Rather Have a Chromebook. *Macworld*, January 2016. (https://www.macworld.com/article/227189/of-cars-trucks-ipads-and-laptops.html) (accessed October 2023).

Spurling, N. (2021) Matters of Time: Materiality and the Changing Temporal Organisation of Everyday Energy Consumption. *Journal of Consumer Culture* 21(2): 146–163.

Standage, T. (1999) *The Victorian Internet: The Remarkable Story of the Telegraph and the Nineteenth Century's On-Line Pioneers*. New York: Berkley Books.

Stankievech, C. (2007) Headphones, Epoche, and L'extimité: A Phenomenology of Interiority. *Offscreen* 11(8–9).

Starosielski, N. (2015) *The Undersea Network*. Durham, NC: Duke University Press.

Steiner, H. and Veel, K. (2017) Negotiating the Boundaries of the Home: The Making and Breaking of Lived and Imagined Walls. *Home Cultures* 14(1): 1–5.

Steiner, H. and Veel, K. (2020) *Tower to Tower: Gigantism in Architecture and Digital Culture*. Cambridge, MA: MIT Press.

Stewart, K. (2011) Atmospheric Attunements. *Environment and Planning D: Society and Space* 29(3): 445–453.

Ssorin-Chaikov, N. (2013) Ethnographic Conceptualism. *Laboratorium* 5(2): 5–18.

Star, S.L. (1999) The Ethnography of Infrastructure. *American Behavioral Scientist* 43(3): 377–391.

Star, S.L. and Ruhleder, K. (1996) Steps Toward an Ecology of Infrastructure: Design and Access for Large Information Spaces. *Information Systems Research* 7(1): 111–134.

Starlink Mission (2022) https://www.spacex.com/launches/sl3-3/ (accessed October 2023).

Strang, V., Edensor, T. and Puckering, J. (Eds) (2018) *From the Lighthouse: Interdisciplinary Reflections on Light. 1st edition.* London: Routledge. https://www.taylorfrancis.com/books/9781317131625 (accessed 21 January 2022).

Strengers, Y. (2013) *Smart Energy Technologies in Everyday Life: Smart Utopia?* Basingstoke: Palgrave MacMillan.

Sumartojo, S. and Pink, S. (2020) *Atmospheres and the Experiential World: Theory and Methods.* London: Routledge.

Sveriges radio (2012) *Isskulptur gav eko i världen.* 12 April. https://sverigesra dio.se/artikel/5053359 (accessed December 2022).

Swish website (n.d.) (https://www.swish.nu/newsroom/stories/quick-commerce-more-than-fast-delivieries) (accessed October 2022).

Sydsvenskan (2021) Länsstyrelsen: Microsofts utsläpp drabbar grannarna. 11 December. https://www.sydsvenskan.se/2021-12-11/lansstyrelsen-microsofts-utslapp-drabbar-grannarna (accessed October 2023).

Sydsvenskan (2022) Mystiskt sken över Skåne var raketbränsle i solljus. 13 August 13 2022. https://www.sydsvenskan.se/2022-08-13/mystiskt-sken-over-skane-var-raketbransle-i-solljus (accessed March 2023).

Telenor Change (n.d.) https://www.telenor.se/handla/tjanster/mobil/cha nge/ (accessed June 2023).

'Ten Grand Goldie' (Official video) (2020) Einstürzende Neubauten. https://www.youtube.com/watch?v=UIjTHkN21Zo (accessed May 2023).

The Spheres (n.d.) https://www.seattlespheres.com (accessed May 2023).

Taylor, C. (2003) *Modern Social Imaginaries.* Durham, NC: Duke University Press.

The World Bank (n.d.) Practitioner's guide: tokenization. https://id4d. worldbank.org/guide/tokenization (accessed March 2023).

Thrift, N. (2007) *Non-Representational Theory: Space, Politics, Affect.* Oxford: Routledge.

Trentmann, F. (2009) Disruption is Normal: Blackouts, Breakdowns and the Elasticity of Everyday Life. In E. Shove, F. Trentmann and R. Wilk (Eds) *Time, Consumption and Everyday Life. Practice, Materiality and Culture* (pp 67–84). Oxford: Berg.

Turkle, S. (2007) *Evocative Objects: Things We Think With.* Cambridge, MA: MIT Press.

Turner, F. (2018) Millenarian Tinkering: The Puritan Roots of the Maker Movement. *Technology and Culture* 59(4): S160–S182.

UNESCO (n.d.) Grimeton Radio Station, Varberg. https://whc.unesco.org/en/list/1134/) (accessed August 2023).

van Lente, H., Spitters, C. and Peine A. (2013) Comparing Technological Hype Cycles: Towards a Theory. *Technological Forecasting & Social Change* 80: 1615–1628.

Vannini, P. (2015) Non-representational Research Methodologies. An Introduction. In P. Vannini (Ed), *Non-representational Methodologies: Re-envisioning Research*. London: Routledge.

Velkova, J., and Kaun, A. (2021) Algorithmic Resistance: Media Practices and the Politics of Repair. *Information, Communication & Society* 24(4): 523–540.

Vermona Retroverb Lancet (n.d.) https://www.vermona.com/en/products/effects/product/retroverb-lancet/ (accessed August 2023).

Vonderau, A. (2018) Scaling the Cloud: Making State and Infrastructure in Sweden. *Ethnos* 84(4): 1–21.

Warde, A. (2021) Review of: The Society of Singularities by Andreas Reckwitz. *British Journal of Sociology* 72(2): 466–469.

Warren, T. (2016) Wanna Be Startin' Somethin': A History of the Windows Start Menu. *The Verge*. https://www.theverge.com/2016/2/11/10923808/microsoft-windows-start-menu-20-years-visual-history (accessed May 2022).

Weber, H. (2010) Head Cocoons: A Sensori-Social History of Earphone Use in West Germany, 1950–2010. *The Senses and Society* 5(3): 339–363.

Weiser, M. (1991) The Computer for the 21st Century. *Scientific American* 265(3): 94–104.

Weiser, M. (1994) The World Is Not a Desktop. *Interactions* (January): 7–8.

Weiser, M., Gold, R. and Brown J.S. (1999) The Origins of Ubiquitous Computing Research at PARC in the Late 1980s. *IBM Systems Journal* 38(4): 693–696.

Wikipedia, 1984 (advertisement) (n.d.) https://en.wikipedia.org/wiki/1984_(advertisement (accessed May 2023).

Wikipedia, early adopter (n.d.) https://en.wikipedia.org/wiki/Early_adopter (accessed May 2023).

Wikipedia, handshake (computing) (n.d.) https://en.wikipedia.org/wiki/Handshake_(computing) (accessed May 2023).

Wikipedia, skeuomorph (n.d.) https://en.wikipedia.org/wiki/Skeuomorph (accessed May 2023).

Wilk, R. (2011) Reflections on Orderly and Disorderly Ethnography. *Ethnologia Europaea. Journal of European Ethnology* 41(1): 15–25.

Willim, R. (1999) Semi-Detached. Computers and The Aesthetic of Ephemerality. In: S. Lundin and L. Åkesson (Eds) *Amalgamations. Fusing Technology and Culture* (pp 18–39). Lund: Nordic Academic Press.

Willim, R. (2002) *Framtid.nu – Flyt och friktion i ett snabbt företag*. Stockholm/ Stehag: Symposion.

Willim, R. (2003a) Claiming the Future: Speed, Business Rhetoric and Computer Practice in a Swedish IT Company. In: C. Garsten and H. Wulff (Eds) *New Technologies at Work: People, Screens and Social Virtuality*. Oxford: Berg.

Willim, R. (2003b) Tools for the Electronic Frontier: Artefacts and Associations in IT Business. *Ethnologia Scandinavica. A Journal for Nordic Ethnology* 33: 21–29.

Willim, R. (2005a) Looking With New Eyes on The Old Factory: On The Rise of Industrial Cool. In: T. O'Dell and P. Billingm (Eds) *Experiencescapes. Tourism, Culture and Economy*. Copenhagen: Copenhagen Business School Press.

Willim, R. (2005b) It's in The Mix – Configuring Industrial Cool. In: O. Löfgren and R. Willim (Eds) *Magic, Culture and the New Economy*. Oxford: Berg.

Willim, R. (2007) *Menuing*: Ethnologia Europaea. *Journal of European Ethnology* 35(1–2): 125–129.

Willim, R. (2013a) Enhancement or Distortion? From the Claude Glass to Instagram. In: *Sarai Reader 09: Projections*: 353–359. Delhi: The Sarai Programme, CSDS.

Willim, R. (2013b) Out of Hand: Reflections on Elsewhereness. In: A. Schneider and C. Wright (Eds) *Anthropology and Art Practice*. Oxford: Bloomsbury.

Willim, R. (2017a) Imperfect Imaginaries. Digitisation, Mundanisation and the Ungraspable. In: G. Koch (Ed) *Digitisation: Theories and Concepts for Empirical Cultural Research* (pp 53–77). London: Routledge.

Willim, R. (2017b) Evoking Imaginaries: Art Probing, Ethnography and More-than-academic Practice. *Sociological Research Online* 22(3): 1–24.

Willim, R. (2023) Probing Mundania: Using Art and Cultural Analysis to Explore Emerging Technologies. *Cultural Analysis* 21(1): 22–37.

Wuthoff, G. (2016) *The Perversity of Things. Hugo Gernsback on Media, Tinkering, and Scientifiction*. Minneapolis, MN: University of Minnesota Press.

ZKM (n.d.) Hugo Gernsback. https://zkm.de/en/person/hugo-gernsback (accessed May 2023).

Zuboff, S. (2019) *The Age of Surveillance Capitalism: The Fight for a Human Future at the New Frontier of Power*. London: Profile Books.

Index